I0489326

Mathuranathan Viswanathan

Digital Modulations using Matlab

Build Simulation Models from Scratch

May 2019

mathuranathan@gmail.com

https://www.gaussianwaves.com

ISBN: 9781521493885

First published June 2017

Revised edition May 2019

The author has used his best endeavours to ensure that the URLs for external websites referred in this book are correct and active at the time of publishing. However, the author bears no responsibility for the referred websites, and can make no guarantee that a site will remain live or that the content is or will remain appropriate.

Dedicated to *Advaith*

Preface

There exist many textbooks that provide an in-depth treatment of various topics in digital modulation techniques. Most of them underscore different theoretical aspects of design and performance analysis of digital modulation techniques. Only a handful of books provide insight on how these techniques can be modeled and simulated. Predominantly, such books utilize the sophisticated built-in functions or toolboxes that are already available in software like Matlab. These built-in functions or toolboxes hide a lot of background computations from the user thereby making it difficult, especially for a learner, to understand how certain techniques are actually implemented inside those functions.

In this book, I intend to show how the theoretical aspects of a digital modulation-demodulation system can be translated into simulation models, using elementary matrix operations in Matlab. Most of the simulation models shown in this book, will not use any of the inbuilt communication toolbox functions. This provides an opportunity for a practicing engineer to understand the basic implementation aspects of modeling various building blocks of a digital modulation system. I intend the book to be used primarily by undergraduate and graduate students in electrical engineering discipline, who wish to learn the basic implementation aspects of a modulation demodulation technique. I assume that the reader has a fair understanding on the fundamentals of programming in Matlab. Readers may consult other textbooks and documentations that cover those topics.

Theoretical aspects of digital modulation techniques will be kept brief. For each topic discussed, a short theoretical background is provided along with the implementation details in the form of Matlab scripts. The Matlab scripts carry inline comments intended to help the reader understand the flow of implementation.

As for the topics concerned, only the basic techniques of modulation and demodulation of various digital modulation techniques are covered. Waveform simulation technique and the complex equivalent baseband simulation model will be provided on a case-by-case basis. Performance simulations of well known digital modulation techniques are also provided. Additionally, simulation and performance of receiver impairments are also provided in a separate chapter.

Chapter 1 introduces some of the basic signal processing concepts that will be used throughout this book. Concepts covered in this chapter include- signal generation techniques for generating well known test signals, interpreting FFT results and extracting magnitude/phase information using FFT, computation of power and energy of a signal, various methods to compute convolution of two signals. Chapter 2 covers the waveform simulation technique for the following digital modulations: BPSK, differentially encoded BPSK, differential BPSK, QPSK, offset QPSK, pi/4 QPSK, CPM and MSK, GMSK, FSK. Power spectral density (PSD) and performance analysis for these techniques are also provided. Chapter 3 covers the complex baseband equivalent models for techniques such as M-ary PAM, M-ary PSK, M-ary QAM and M-ary FSK modulations. Chapter 4 covers the performance simulation using the models built in Chapter 3. Chapter 5 covers the aspects of using various linear equalizers in a simple communication link. Design and implementation of two important types of equalizers namely the zero-forcing equalizer and the MMSE equalizer are covered. Chapter 6 covers the topic of modeling receiver impairment, estimation and compensation for such impairments and a sample performance simulation for such case. Reference texts are cited in square brackets within the text and the references are provided at the end of each chapter.

The documentation for the Matlab scripts shown in this book, is available at the following URL: `https://www.gaussianwaves.com/digital_modulations_matlab/doc`

The scripts are thoroughly checked for integrity and they will execute without any error. If you face any issues during execution, do not hesitate to provide feedback or contact me via the email provided below. Alternatively, technical queries can be posted in the following forum: `https://www.gaussianwaves.com/discuss/`

Finally, it is a pleasure to acknowledge the help I received while writing this book. I thank my wife *Varsha Mathuranathan* for getting the manuscript edited so quickly and for her support during the review process that greatly helped improve the manuscript. I also thank the numerous reviewers for their generous comments that helped improve the contents of this book.

Singapore,   *Mathuranathan Viswanathan*
June 2017   *mathuranathan@gmail.com*

Contents

Chapter 1
Essentials of Signal Processing

Abstract This chapter introduces some of the basic signal processing concepts that will be used throughout this book. The goal is to enable the reader to appreciate the concepts and apply them in building a basic communication system. Concepts covered include - signal generation techniques for generating well known test signals like rectangular pulse, sine wave, square wave, chirp signal and gaussian pulse, interpreting FFT results and extracting magnitude/phase information using FFT, computation of power and energy of a signal, various methods to compute convolution of two signals, analytic signal and applications, FIR/IIR filters.

1.1 Generating standard test signals

In experimental modeling and simulation, simple test inputs such as sinusoidal, rectangular pulse, gaussian pulse, and chirp signals are widely used. These test signals act as stimuli for the simulation model and the response of the model to the stimuli is of great interest in design verification.

1.1.1 Sinusoidal signals

In order to generate a sine wave, the first step is to fix the frequency f of the sine wave. For example, we wish to generate a $f = 10Hz$ sine wave whose minimum and maximum amplitudes are $-1V$ and $+1V$ respectively. Given the frequency of the sinewave, the next step is to determine the sampling rate.

For baseband signals, the sampling is straight forward. By *Nyquist Shannon sampling theorem*, for faithful reproduction of a continuous signal in discrete domain, one has to sample the signal at a rate f_s higher than at-least twice the maximum frequency f_m contained in the signal (actually, it is twice the one-sided bandwidth occupied by a real signal. For a baseband signal bandwidth (0 to f_m) and maximum frequency f_m in a given band are equivalent).

Matlab is a software that processes everything in digital. In order to obtain a smooth sine wave, the sampling rate must be far higher than the prescribed minimum required sampling rate which is at least twice the frequency f - as per *Nyquist-Shannon theorem*. Hence we need to sample the input signal at a rate significantly higher than what the Nyquist criterion dictates. Higher oversampling rate requires more memory for signal storage. It is advisable to keep the oversampling factor to an acceptable value.

An oversampling factor of 30 is chosen in the following code snippet. This is to plot a smooth continuous-like sine wave. Thus the sampling rate becomes $f_s = 30 \times f = 30 \times 10 = 300Hz$. If a phase shift is desired for the sine wave, specify it too. The resulting plot from the code snippet shown next, is given in Figure 1.1.

Program 1.1: *sinusoidal_signal.m*: Simulate a sinusoidal signal with given sampling rate

```
1  f=10; %frequency of sine wave
2  overSampRate=30; %oversampling rate
3  fs=overSampRate*f; %sampling frequency
4  phase = 1/3*pi; %desired phase shift in radians
5  nCyl = 5; %to generate five cycles of sine wave
6  t=0:1/fs:nCyl*1/f-1/fs; %time base
7
8  g=sin(2*pi*f*t+phase); %replace with cos if a cosine wave is desired
9  plot(t,g); title(['Sine Wave f=', num2str(f), 'Hz']);
```

Fig. 1.1: A $10Hz$ sinusoidal wave with 5 cycles and phase shift $1/3\pi$ radians

1.1.2 Square wave

The most logical way of transmitting information across a communication channel is through a stream of square pulse – a distinct pulse for '0' and another for '1'. Digital signals are graphically represented as square waves with certain symbol/bit period. Square waves are also used universally in switching circuits, as clock signals synchronizing various blocks of digital circuits, as reference clock for a given system domain and so on.

Square wave manifests itself as a wide range of harmonics in frequency domain and therefore can cause electromagnetic interference. Square waves are periodic and contain odd harmonics when expanded as Fourier Series (where as signals like saw-tooth and other real word signals contain harmonics at all integer frequencies). Since a square wave literally expands to infinite number of odd harmonic terms in frequency domain, approximation of square wave is another area of interest. The number of terms of its Fourier Series expansion, taken for approximating the square wave is often seen as *Gibbs phenomenon*, which manifests as ringing effect at the corners of the square wave in time domain.

True square waves are a special class of rectangular waves with 50% duty cycle. Varying the duty cycle of a rectangular wave leads to pulse width modulation, where the information is conveyed by changing the duty-cycle of each transmitted rectangular wave. A true square wave can be simply generated by applying *signum function* over a periodic function.

$$g(t) = sgn\left[sin(2\pi ft)\right] \tag{1.1}$$

where f is the desired frequency of the square wave and the signum function is defined as

$$sgn(x) = \begin{cases} -1 & if\ x < 0, \\ 0 & if\ x = 0, \\ 1 & if\ x > 0 \end{cases} \tag{1.2}$$

Program 1.2: *square_wave.m*: Generate a square wave with given sampling rate

```
f=10; %frequency of sine wave in Hz
overSampRate=30; %oversampling rate
fs=overSampRate*f; %sampling frequency
nCyl = 5; %to generate five cycles of square wave
t=0:1/fs:nCyl*1/f-1/fs; %time base
g = sign(sin(2*pi*f*t));
%g=square(2*pi*f*t,50);%inbuilt fn:(signal proc toolbox)
plot(t,g); title(['Square Wave f=', num2str(f), 'Hz']);
```

Fig. 1.2: A $10Hz$ square wave with 5 cycles and $50-50$ duty cycle

1.1.3 Rectangular pulse

An isolated rectangular pulse of amplitude A and duration T is represented mathematically as

$$g(t) = A \cdot rect\left(\frac{t}{T}\right) \tag{1.3}$$

where,

$$rect(t) = \begin{cases} 1 & if \ |t| < \frac{1}{2} \\ \frac{1}{2} & if \ |t| = \frac{1}{2} \\ 0 & if \ |t| > \frac{1}{2} \end{cases} \tag{1.4}$$

The following code simulates a rectangular pulse with desired pulse width and the resulting plot is shown in Figure 1.3.

Program 1.3: *rectangular_pulse.m*: Generating a rectangular pulse with desired pulse width

```
fs=500; %sampling frequency
T=0.2; %width of the rectangule pulse in seconds
t=-0.5:1/fs:0.5; %time base
g=(t >-T/2) .* (t<T/2) + 0.5*(t==T/2) + 0.5*(t==-T/2);
%g=rectpuls(t,T); %using inbuilt function (signal proc toolbox)
plot(t,g);title(['Rectangular Pulse width=', num2str(T),'s']);
```

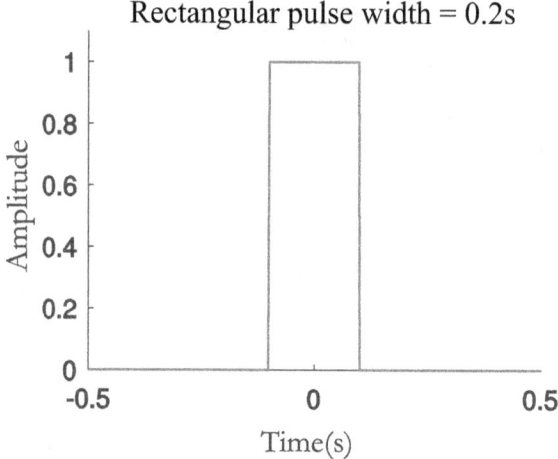

Fig. 1.3: A rectangular pulse having pulse-width 0.2*s*

1.1.4 Gaussian pulse

In digital communications, Gaussian Filters are employed in Gaussian Minimum Shift Keying - GMSK (see section 2.11) and Gaussian Frequency Shift Keying (GFSK). Two dimensional Gaussian Filters are used in

Image processing to produce Gaussian blurs. The impulse response of a Gaussian Filter is written as a Gaussian function as follows

$$g(t) = \frac{1}{\sqrt{2\pi}\sigma} e^{-\frac{t^2}{2\sigma^2}} \tag{1.5}$$

The following code generates a Gaussian Pulse with $\sigma = 0.1s$. The resulting plot is given in Figure 1.4

Program 1.4: *gaussian_pulse.m*: Generating a Gaussian pulse with desired pulse width

```
1  fs=80; %sampling frequency
2  sigma=0.1;%standard deviation
3  t=-0.5:1/fs:0.5; %time base
4  g=1/(sqrt(2*pi)*sigma)*(exp(-t.^2/(2*sigma^2)));
5  plot(t,g); title(['Gaussian Pulse \sigma=', num2str(sigma),'s']);
```

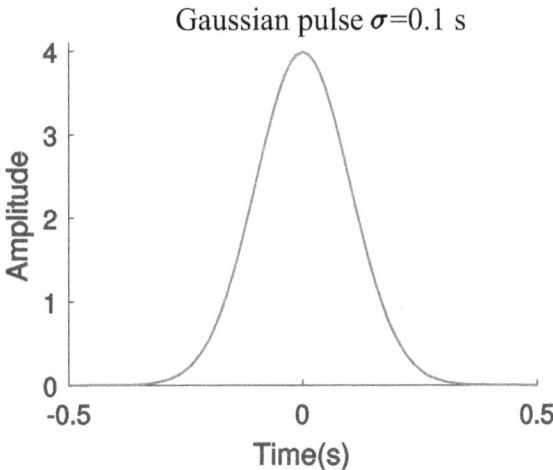

Fig. 1.4: A Gaussian pulse with $\sigma = 0.1s$

1.1.5 Chirp signal

All the signals discussed so far do not change in frequency over time. Obtaining a signal with time-varying frequency is of main focus here. A signal that varies in frequency over time is called *chirp*. The frequency of the chirp signal can vary from low to high frequency (up-chirp) or from high to low frequency (low-chirp).

Chirp signals are encountered in many applications ranging from radar, sonar, spread spectrum, optical communication, image processing, doppler effect, motion of a pendulum, as gravitation waves, manifestation as Frequency Modulation (FM), echo location etc.

A linear chirp signal sweeps the frequency from low to high frequency (or vice-versa) linearly. One approach to generate a chirp signal is to concatenate a series of segments of sine waves each with increasing(or decreasing) frequency in order. This method introduces discontinuities in the chirp signal due to the mismatch in the phases of each such segments. Modifying the equation of a sinusoid to generate a chirp signal is a better approach.

The equation for generating a sinusoidal (cosine here) signal with amplitude A, angular frequency ω_0 and initial phase ϕ is

$$x(t) = A\,cos(\omega_0 t + \phi) \tag{1.6}$$

This can be written as a function of instantaneous phase

$$x(t) = A\,cos[\theta(t)] \tag{1.7}$$

where $\theta(t) = \omega_0(t) + \phi$ is the instantaneous phase of the sinusoid and it is linear in time. The time derivative of instantaneous phase $\theta(t)$, is equal to the angular frequency ω of the sinusoid.

$$\omega(t) = \frac{d}{dt}\theta(t) \tag{1.8}$$

Instead of having the phase linear in time, let's change the phase to quadratic form and thus render it non-linear. For some constant α,

$$\theta(t) = 2\pi\alpha t^2 + 2\pi f_0 t + \phi \tag{1.9}$$

The first derivative of the phase is the instantaneous angular frequency as given by

$$\omega_i(t) = \frac{d}{dt}\theta(t) = 4\pi\alpha t + 2\pi f_0 \tag{1.10}$$

Hence, the time-varying frequency in Hz is given by

$$f_i(t) = 2\alpha t + f_0 \tag{1.11}$$

In the above equation, the frequency is no longer a constant, rather it is of time-varying nature with initial frequency given by f_0. Thus, from the above equation, given a time duration T, the rate of change of frequency is given by

$$k = 2\alpha = \frac{f_1 - f_0}{T} \tag{1.12}$$

where, f_0 is the starting frequency of the sweep, f_1 is the frequency at the end of the duration T. Substituting equations 1.11 and 1.12 in 1.10,

$$\omega_i(t) = \frac{d}{dt}\theta(t) = 2\pi\,(kt + f_0) \tag{1.13}$$

From equations 1.8 and 1.13

$$\theta(t) = \int \omega_i(t)\,dt$$

$$= 2\pi \int (kt + f_0)\,dt = 2\pi \left(k\frac{t^2}{2} + f_0 t \right) + \phi_0$$

$$= 2\pi \left(k\frac{t^2}{2} + f_0 t \right) + \phi_0 = 2\pi \left(\frac{k}{2}t + f_0 \right) t + \phi_0 \tag{1.14}$$

where, ϕ_0 is a constant which will act as the initial phase of the sweep. Thus the modified equation for generating a chirp signal is given by

$$x(t) = A\,cos\left[2\pi f(t)\,t + \phi_0\right] \tag{1.15}$$

where, the time-varying frequency function is given by

$$f(t) = \frac{k}{2}t + f_0 \tag{1.16}$$

A chirp signal can be generated without using the inbuilt `chirp` function in Matlab's signal processing toolbox. This is done by implementing a function that utilized equations 1.12, 1.15 and 1.16. The implemented function is shown next.

The implemented function accepts five arguments to generate a chirp signal: t - the discrete time base for generating the chirp, f_0 - the initial frequency of the sweep, f_1 - the frequency at time t_1 and ϕ_0 - the initial phase of the chirp which is an optional argument. If the initial phase is not supplied to the function, it assumes the initial phase to be zero.

Program 1.5: *chirp_signal.m*: Function to generate a Chirp signal

```
function g=chirp_signal(t,f0,t1,f1,phase)
%g = chirp_signal(t,f0,t1,f1) generates samples of a linearly
%swept-frequency signal at the time instances defined in timebase
%array t. The instantaneous frequency at time 0 is f0 Hertz.
%The instantaneous frequency f1 is achieved at time t1. The argument
%'phase' is optional. It defines the initial phase of the signal
%defined in radians. By default phase=0 radian
if nargin==4, phase=0; end
t0=t(1); T=t1-t0; k=(f1-f0)/T;
g=cos(2*pi*(k/2*t+f0).*t+phase);
end
```

The following test script utilizes the above function and generates a chirp with starting frequency $f_0 = 1Hz$ at the start of the time base and $f_1 = 25Hz$ at $t_1 = 1s$ which is the end of the time base. From the PSD plot, it can be ascertained that the signal energy is concentrated only upto $25Hz$. The resulting plot is shown in Figure 1.5.

Program 1.6: *chirp_test.m*: Generating and plotting a chirp signal

```
fs=500; %sampling frequency
t=0:1/fs:1; %time base - upto 1 second
f0=1;% starting frequency of the chirp
f1=fs/20; %frequency of the chirp at t1=1 second
g = chirp_signal(t,f0,1,f1);
plot(t,g); title('Chirp Signal');
```

1.2 Interpreting FFT results - complex DFT, frequency bins and FFTShift

Often, one is confronted with the problem of converting a time domain signal to frequency domain and vice-versa. Fourier Transform is an excellent tool to achieve this conversion and is ubiquitously used in many applications. In signal processing , a time domain signal can be *continuous* or *discrete* and it can be *aperiodic* or *periodic*. This gives rise to four types of Fourier transforms as listed in Table 1.1.

From Table 1.1, we note that when the signal is discrete in one domain, it will be periodic in other domain. Similarly, if the signal is continuous in one domain, it will be aperiodic (non-periodic) in another domain. For simplicity, lets not venture into the specific equations for each of the transforms above. We will limit our discussion to DFT, that is widely available as part of software packages like Matlab, Scipy(python) etc.., however we can approximate other transforms using DFT.

Fig. 1.5: A finite record of a chirp signal

Table 1.1: Four types of Fourier Transforms

Transform	Nature of time domain signal	Nature of frequency spectrum
Fourier Transform (FT), (a.k.a Continuous Time Fourier Transform (CTFT))	continuous, non-periodic	non-periodic,continuous
Discrete-time Fourier Transform (DTFT)	discrete, non-periodic	periodic,continuous
Fourier Series (FS)	continuous, periodic	non-periodic, discrete
Discrete Fourier Transform (DFT)	discrete, periodic	periodic,discrete

1.2.1 Real and complex DFT

For each of the listed transforms above, there exist a real version and complex version. The real version of the transform, takes in a real numbers and gives two sets of real frequency domain points - one set representing coefficients over *cosine* basis function and the other set representing the coefficient over *sine* basis function. The complex version of the transforms represent positive and negative frequencies in a single array. The complex versions are flexible that it can process both complex valued signals and real valued signals. Figure 1.6 captures the difference between real DFT and complex DFT.

1.2.1.1 Real DFT

Consider the case of N-point *real* DFT , it takes in N samples of *real-valued* time domain waveform $x[n]$ and gives two arrays of length $N/2+1$ each set projected on cosine and sine functions respectively.

$$X_{re}[k] = \frac{2}{N} \sum_{n=0}^{N-1} x[n] cos\left(\frac{2\pi kn}{N}\right)$$

$$X_{im}[k] = -\frac{2}{N} \sum_{n=0}^{N-1} x[n] sin\left(\frac{2\pi kn}{N}\right)$$

(1.17)

Here, the time domain index n runs from $0 \rightarrow N$, the frequency domain index k runs from $0 \rightarrow N/2$ The real-valued time domain signal $x[n]$ can be synthesized from the real DFT pairs as

$$x[n] = \sum_{k=0}^{N/2} X_{re}[K]cos\left(\frac{2\pi kn}{N}\right) - X_{im}[K]sin\left(\frac{2\pi kn}{N}\right) \tag{1.18}$$

Caveat: When using the synthesis equation, the values $X_{re}[0]$ and $X_{re}[N/2]$ must be divided by two. This problem is due to the fact that we restrict the analysis to real-values only. These type of problems can be avoided by using complex version of DFT.

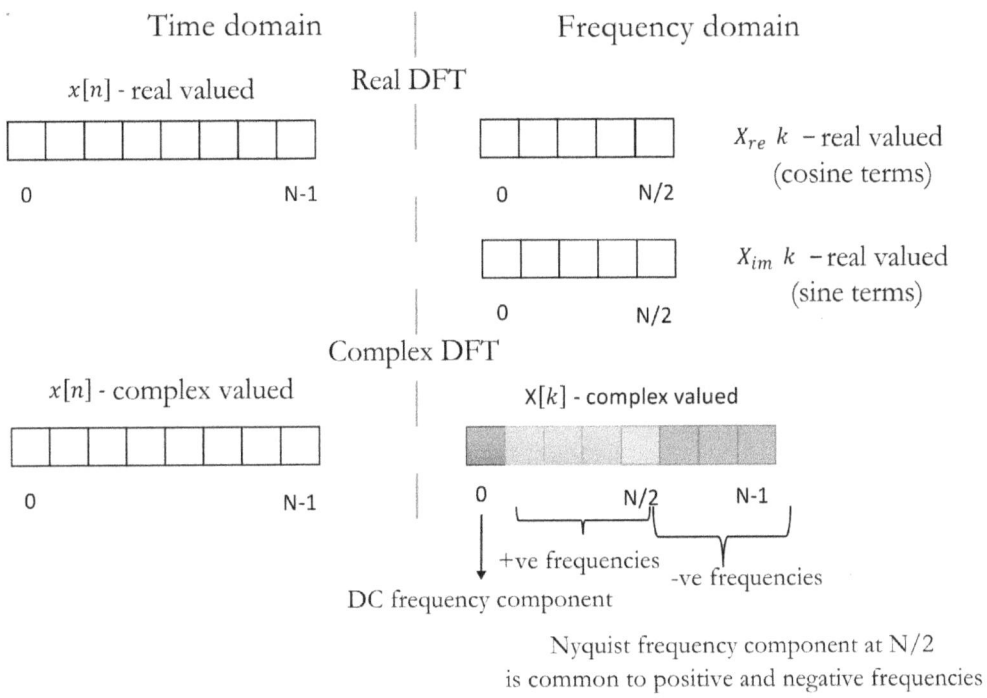

Fig. 1.6: Real and complex DFT

1.2.1.2 Complex DFT

Consider the case of N-point *complex* DFT, it takes in N samples of *complex-valued* time domain waveform $x[n]$ and produces an array $X[k]$ of length N.

$$X[k] = \frac{1}{N} \sum_{n=0}^{N-1} x[n]e^{-j2\pi kn/N} \tag{1.19}$$

The arrays values are interpreted as follows

- $X[0]$ represents DC frequency component
- Next $N/2$ terms are positive frequency components with $X[N/2]$ being the Nyquist frequency (which is equal to half of sampling frequency)

- Next $N/2 - 1$ terms are negative frequency components (note: negative frequency components are the phasors rotating in opposite direction, they can be optionally omitted depending on the application)

The corresponding synthesis equation (reconstruct $x[n]$ from frequency domain samples $X[k]$) is

$$x[n] = \sum_{k=0}^{N-1} X[k] e^{j2\pi kn/N} \tag{1.20}$$

From these equations we can see that the real DFT is computed by projecting the signal on cosine and sine basis functions. However, the complex DFT projects the input signal on exponential basis functions (Euler's formula - $e^{i\theta} = cos\theta + isin\theta$ connects these two concepts).

When the input signal in the time domain is real valued, the complex DFT zero-fills the imaginary part during computation (That's its flexibility and avoids the caveat needed for real DFT). Figure 1.7 shows how to interpret the raw FFT results in Matlab that computes complex DFT. The specifics will be discussed next with an example.

Fig. 1.7: Interpretation of frequencies in complex DFT output

1.2.2 Fast Fourier Transform (FFT)

The FFT function in Matlab is an algorithm published in 1965 by J.W.Cooley and J.W.Tuckey for efficiently calculating the DFT [1]. It exploits the special structure of DFT when the signal length is a power of 2, when this happens, the computation complexity is significantly reduced. FFT length is generally considered as power of 2 - this is called *radix* $- 2$ FFT which exploits the twiddle factors. The FFT length can be odd as used in this particular FFT implementation - Prime-factor FFT algorithm [2] [3] where the FFT length factors into two co-primes.

FFT is widely available in software packages like Matlab, Scipy etc.., FFT in Matlab/Scipy implements the complex version of DFT. Matlab's FFT implementation computes the complex DFT that is very similar to above equations except for the scaling factor. For comparison, the Matlab's FFT implementation computes the complex DFT and its inverse as

$$X[k] = \sum_{n=0}^{N-1} x[n]e^{-j2\pi kn/N}$$

$$x[n] = \frac{1}{N}\sum_{k=0}^{N-1} X[k]e^{j2\pi kn/N} \qquad (1.21)$$

The Matlab commands that implement the above equations are FFT and IFFT respectively. The corresponding syntax is as follows

```
X = fft(x,N) %compute X[k]
x = ifft(X,N) %compute x[n]
```

1.2.3 Interpreting the FFT results

Let's assume that the $x[n]$ is the time domain cosine signal of frequency $f_c = 10\,Hz$ that is sampled at a frequency $f_s = 32 * fc$ for representing it in the computer memory (Figure 1.8).

Program 1.7: Generating a 2 seconds record of 10 Hz cosine wave

```
1  fc=10;%frequency of the carrier
2  fs=32*fc;%sampling frequency with oversampling factor=32
3  t=0:1/fs:2-1/fs;%2 seconds duration
4  x=cos(2*pi*fc*t);%time domain signal (real number)
5  subplot(3,1,1); plot(t,x);hold on; %plot the signal
6  title('x[n]=cos(2 \pi 10 t)'); xlabel('t=nT_s'); ylabel('x[n]');
```

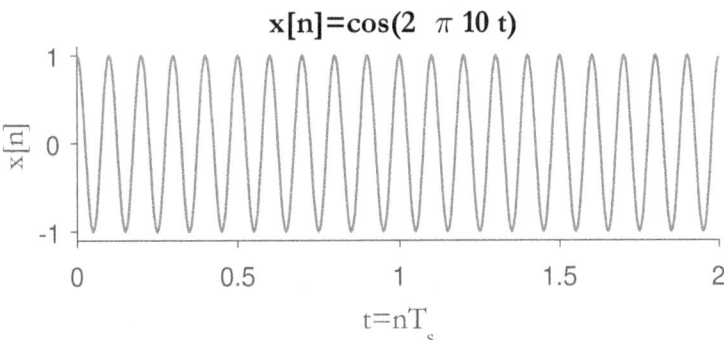

Fig. 1.8: A 2 seconds record of 10 Hz cosine wave

Let's consider taking a $N = 256$ point FFT, which is the 8th power of 2.

Program 1.8: Taking N-point DFT

```
1  N=256; %FFT size
2  X = fft(x,N);%N-point complex DFT, output contains DC at index 1
3  %Nyquist frequency at N/2+1 th index positive frequencies from
4  %index 2 to N/2 negative frequencies from index N/2+1 to N
```

Note: The FFT length should be sufficient to cover the entire length of the input signal. If N is less than the length of the input signal, the input signal will be truncated when computing the FFT. In our case, the cosine wave is of 2 seconds duration and it will have 640 points (a $10Hz$ frequency wave sampled at 32 times oversampling factor will have $2 \times 32 \times 10 = 640$ samples in 2 seconds of the record). Since our input signal is periodic, we can safely use $N = 256$ point FFT, anyways the FFT will extend the signal when computing the FFT. Due to Matlab's index starting at 1, the DC component of the FFT decomposition is present at index 1.

```
>>X(1)
1.1762e-14  (approximately equal to zero)
```

That's pretty easy. Note that the index for the raw FFT are integers from $1 \rightarrow N$. We need to process it to convert these integers to *frequencies*. That is where the *sampling* frequency counts. Each point/bin in the FFT output array is spaced by the frequency resolution Δf, that is calculated as

$$\Delta f = \frac{f_s}{N} \tag{1.22}$$

where, f_s is the sampling frequency and N is the FFT size that is considered. Thus, for our example, each point in the array is spaced by the frequency resolution

$$\Delta f = \frac{f_s}{N} = \frac{32 * f_c}{256} = \frac{320}{256} = 1.25Hz \tag{1.23}$$

The $10Hz$ cosine signal will register a spike at the 8th sample (10/1.25=8) - located at index 9 in Figure 1.9.

```
>> abs(X(8:10)) %display samples 7 to 9
ans = 0.0000 128.0000  0.0000
```

Therefore, from the frequency resolution, the entire frequency axis can be computed as

Program 1.9: Computing the frequency axis for plotting

```
1  %calculate frequency bins with FFT
2  df=fs/N %frequency resolution
3  sampleIndex = 0:N-1; %raw index for FFT plot
4  f=sampleIndex*df; %x-axis index converted to frequencies
```

Now, plot the absolute value of the FFT against frequencies - the resulting plot is shown in the Figure 1.9.

Program 1.10: Plotting the frequency domain view of a signal

```
1  subplot(3,1,2); stem(sampleIndex,abs(X)); %sample values on x-axis
2  title('X[k]'); xlabel('k'); ylabel('|X(k)|');
3  subplot(3,1,3); stem(f,abs(X)); %x-axis represent frequencies
4  title('X[f]'); xlabel('frequencies (f)'); ylabel('|X(f)|');
```

After the frequency axis is properly transformed with respect to the sampling frequency, we note that the cosine signal has registered a spike at $10Hz$. In addition to that, it has also registered a spike at $256 - 8 = 248^{th}$ sample that belongs to negative frequency portion. Since we know the nature of the signal, we can optionally ignore the negative frequencies. The sample at the Nyquist frequency ($f_s/2$) mark the boundary between the positive and negative frequencies.

```
>> nyquistIndex=N/2+1;
>> X(nyquistIndex-2:nyquistIndex+2).'
```

```
ans =
1.0e-13 *
-0.2428 + 0.0404i
-0.1897 + 0.0999i
-0.3784
-0.1897 - 0.0999i
-0.2428 - 0.0404i
```

Note that the complex numbers surrounding the Nyquist index are complex conjugates and they represent positive and negative frequencies respectively.

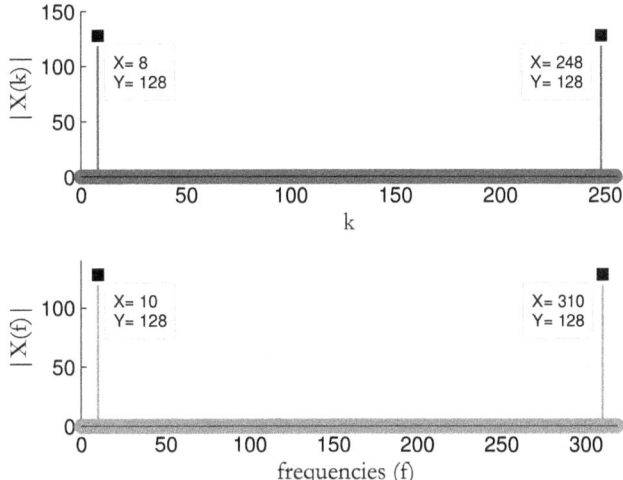

Fig. 1.9: Magnitude response from FFT plotted against - sample index (top) and computed frequencies (bottom)

1.2.4 FFTShift

From Figure 1.9, we see that the frequency axis starts with DC, followed by positive frequency terms which is in turn followed by the negative frequency terms. To introduce proper order in the x-axis, one can use FFTshift function Matlab, which arranges the frequencies in order: negative frequencies $\rightarrow$ DC $\rightarrow$ positive frequencies. The fftshift function need to be carefully used when N is odd.

For even N, the original order returned by FFT is as follows (note: here, all indices corresponds to Matlab's index)

- $X[1]$ represents DC frequency component
- $X[2]$ to $X[N/2]$ terms are positive frequency components
- $X[N/2+1]$ is the Nyquist frequency ($F_s/2$) that is common to both positive and negative frequencies. We will consider it as part of negative frequencies to have the same equivalence with the *fftshift* function
- $X[N/2+1]$ to $X[N]$ terms are considered as negative frequency components

FFTshift shifts the DC component to the center of the spectrum. It is important to remember that the Nyquist frequency at the (N/2+1)th Matlab index is common to both positive and negative frequency sides. FFTshift command puts the Nyquist frequency in the negative frequency side. This is captured in the Figure 1.10.

Fig. 1.10: Role of FFTShift in ordering the frequencies

Therefore, when N is even, ordered frequency axis is set as

$$f = \Delta f \left[-\frac{N}{2} : 1 : \frac{N}{2} - 1 \right] = \frac{f_s}{N} \left[-\frac{N}{2} : 1 : \frac{N}{2} - 1 \right] \qquad (1.24)$$

When N is odd, the ordered frequency axis should be set as

$$f = \Delta f \left[-\frac{N+1}{2} : 1 : \frac{N+1}{2} - 1 \right] = \frac{f_s}{N} \left[-\frac{N+1}{2} : 1 : \frac{N+1}{2} - 1 \right] \qquad (1.25)$$

The following code snippet, computes the fftshift using both the manual method and using the Matlab's in-built command. The results are plotted by superimposing them on each other. The plot in Figure 1.11 shows that both the manual method and fftshift method are in good agreement. Comparing the bottom figures in the Figure 1.9 and Figure 1.11, we see that the ordered frequency axis is more meaningful to interpret.

Program 1.11: Plotting magnitude response using FFTShift

```
1   %two-sided FFT with negative frequencies on left and positive
2   %frequencies on right. Following works only if N is even,
3   %for odd N see equation above
4   X1 = [(X(N/2+1:N)) X(1:N/2)]; %order frequencies without using fftShift
5   X2 = fftshift(X);%order frequencies by using fftshift
6
7   df=fs/N; %frequency resolution
8   sampleIndex = -N/2:N/2-1; %raw index for FFT plot
9   f=sampleIndex*df; %x-axis index converted to frequencies
```

```
10
11  %plot ordered spectrum using the two methods
12  figure;subplot(2,1,1);stem(sampleIndex,abs(X1));hold on;
13  stem(sampleIndex,abs(X2),'r') %sample index on x-axis
14  title('Frequency Domain'); xlabel('k'); ylabel('|X(k)|');
15
16  subplot(2,1,2);stem(f,abs(X1)); stem(f,abs(X2),'r')
17  xlabel('frequencies (f)'); ylabel('|X(f)|');%frequencies on x-axis
```

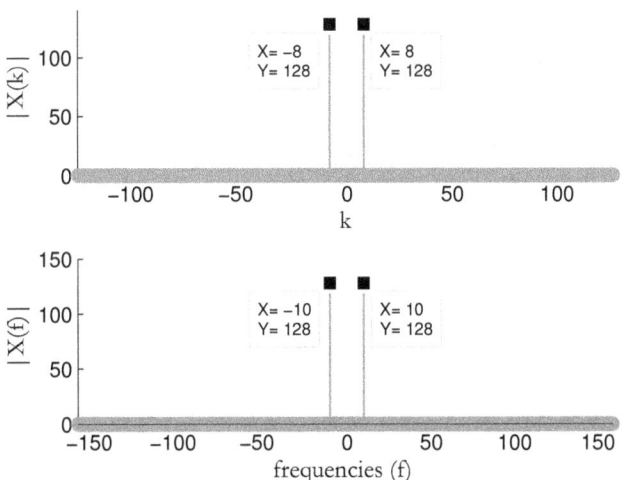

Fig. 1.11: Magnitude response of FFT result after applying FFTShift : plotted against sample index (top) and against computed frequencies (bottom)

1.2.5 IFFTShift

One can undo the effect of fftshift by employing ifftshift function. The ifftshift function restores the raw frequency order. If the FFT output is ordered using fftshift function, then one must restore the frequency components back to original order *before* taking IFFT. Following statements are equivalent.

```
X = fft(x,N) %compute X[k]
x = ifft(X,N) %compute x[n]
```

```
X = fftshift(fft(x,N)); %take FFT and rearrange frequency order
x = ifft(ifftshift(X),N)% restore raw freq order and then take IFFT
```

1.2.6 Some observations on FFTShift and IFFTShift

When N is odd and for an arbitrary sequence, the fftshift and ifftshift functions will produce different results. However, when they are used in tandem, it restores the original sequence.

```
>> x=[0,1,2,3,4,5,6,7,8]
0   1   2   3   4   5   6   7   8
>> fftshift(x)
5   6   7   8   0   1   2   3   4
>> ifftshift(x)
4   5   6   7   8   0   1   2   3
>> ifftshift(fftshift(x))
0   1   2   3   4   5   6   7   8
>> fftshift(ifftshift(x))
0   1   2   3   4   5   6   7   8
```

When N is even and for an arbitrary sequence, the fftshift and ifftshift functions will produce the same result. When they are used in tandem, it restores the original sequence.

```
>> x=[0,1,2,3,4,5,6,7]
0   1   2   3   4   5   6   7
>> fftshift(x)
4   5   6   7   0   1   2   3
>> ifftshift(x)
4   5   6   7   0   1   2   3
>> ifftshift(fftshift(x))
0   1   2   3   4   5   6   7
>> fftshift(ifftshift(x))
0   1   2   3   4   5   6   7
```

1.3 Obtaining magnitude and phase information from FFT

For the discussion here, lets take an arbitrary cosine function of the form $x(t) = A\cos(2\pi f_c t + \phi)$ and proceed step by step as follows

- Represent the signal $x(t)$ in computer (discrete-time) and plot the signal (time domain)
- Represent the signal in frequency domain using FFT ($X[k]$)
- Extract amplitude and phase information from the FFT result
- Reconstruct the time domain signal from the frequency domain samples

1.3.1 Discrete-time domain representation

Consider a cosine signal of amplitude $A = 0.5$, frequency $f_c = 10Hz$ and phase $\phi = \pi/6$ radians (or $30°$)

$$x(t) = 0.5\cos(2\pi 10t + \pi/6) \qquad (1.26)$$

In order to represent the continuous time signal $x(t)$ in computer memory (Figure 1.12), we need to sample the signal at sufficiently high rate (according to Nyquist sampling theorem). I have chosen a oversampling factor of 32 so that the sampling frequency will be $f_s = 32 \times f_c$, and that gives 640 samples in a 2 seconds duration of the waveform record.

Program 1.12: Representing and storing a signal in computer memory

```
1  A = 0.5; %amplitude of the cosine wave
2  fc=10;%frequency of the cosine wave
3  phase=30; %desired phase shift of the cosine in degrees
4  fs=32*fc;%sampling frequency with oversampling factor 32
5  t=0:1/fs:2-1/fs;%2 seconds duration
6
7  phi = phase*pi/180; %convert phase shift in degrees in radians
8  x=A*cos(2*pi*fc*t+phi);%time domain signal with phase shift
9  figure; plot(t,x); %plot the signal
```

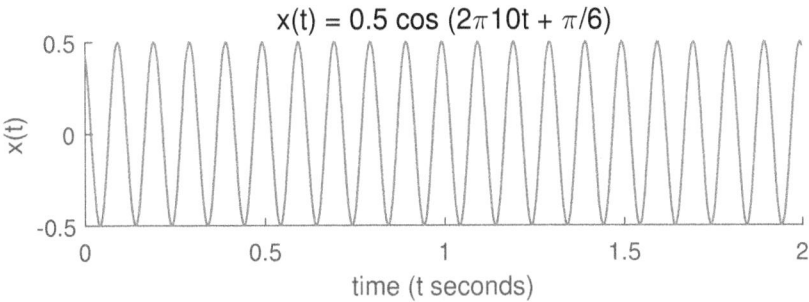

Fig. 1.12: Finite record of a cosine signal

1.3.2 Representing the signal in frequency domain using FFT

Let's represent the signal in frequency domain using the FFT function. The FFT function computes N-point complex DFT. The length of the transformation N should cover the signal of interest otherwise we will some loose valuable information in the conversion process to frequency domain. However, we can choose a reasonable length if we know about the nature of the signal. For example, the cosine signal of our interest is periodic in nature and is of length 640 samples (for 2 seconds duration signal). We can simply use a lower number $N = 256$ for computing the FFT. In this case, only the first 256 time domain samples will be considered for taking FFT. However, we do not need to worry about loss of any valuable information, as the 256 samples will have sufficient number of cycles to extract the frequency of the signal.

Program 1.13: Represent the signal in frequency domain using FFT

```
1  N=256; %FFT size
2  X = 1/N*fftshift(fft(x,N));%N-point complex DFT
```

In the code above, `fftshift` is used only for obtaining a nice double-sided frequency spectrum that delineates negative frequencies and positive frequencies in order. This transformation is not necessary. A scaling factor $1/N$ was used to account for the difference between the FFT implementation in Matlab and the text definition of complex DFT as given in equation 1.19.

1.3.2.1 Extract amplitude of frequency components (amplitude spectrum)

The FFT function computes the complex DFT and the hence the results in a sequence of complex numbers of form $X_{re} + jX_{im}$. The amplitude spectrum is obtained

$$|X[k]| = \sqrt{X_{re}^2 + X_{im}^2} \tag{1.27}$$

For obtaining a double-sided plot, the ordered frequency axis, obtained using `fftshift`, is computed based on the sampling frequency and the amplitude spectrum is plotted (Figure 1.13).

Program 1.14: Extract amplitude information from the FFT result

```
df=fs/N; %frequency resolution
sampleIndex = -N/2:N/2-1; %ordered index for FFT plot
f=sampleIndex*df; %x-axis index converted to ordered frequencies
stem(f,abs(X)); %magnitudes vs frequencies
xlabel('f (Hz)'); ylabel('|X(k)|');
```

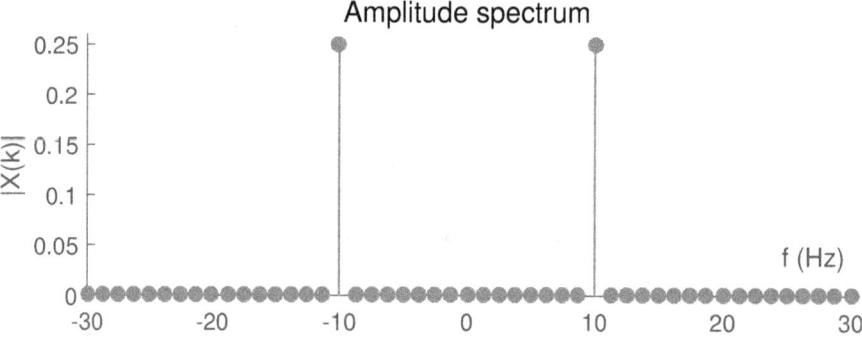

Fig. 1.13: Extracted amplitude information from the FFT result

1.3.2.2 Extract phase of frequency components (phase spectrum)

Extracting the correct phase spectrum is a tricky business. I will show you why it is so. The phase of the spectral components are computed as

$$\angle X[k] = tan^{-1} \left(\frac{X_{im}}{X_{re}} \right) \tag{1.28}$$

The equation 1.28 looks naive, but one should be careful when computing the inverse tangents using computers. The obvious choice for implementation seems to be the *atan* function in Matlab. However, usage of *atan* function will prove disastrous unless additional precautions are taken. The *atan* function computes the inverse tangent over two quadrants only, i.e, it will return values only in the $[-\pi/2, \pi/2]$ interval. Therefore, the phase need to be unwrapped properly. We can simply fix this issue by computing the inverse tangent over all the four quadrants using the $atan2(X_{img}, X_{re})$ function. Let's compute and plot the phase information using *atan2* function and see how the phase spectrum looks.

Program 1.15: Extracting phase information from the FFT result using atan2 function

```
1  phase=atan2(imag(X),real(X))*180/pi; %phase information
2  plot(f,phase); %phase vs frequencies
```

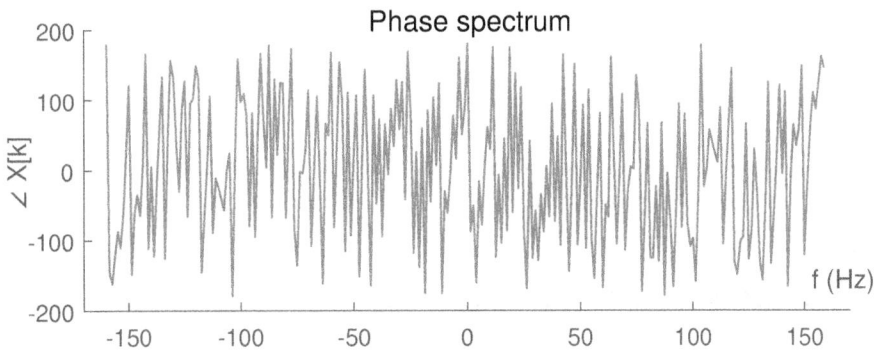

Fig. 1.14: Extracted phase information from the FFT result - phase spectrum is noisy

The phase spectrum in Figure 1.14 is completely noisy, which is unexpected. The phase spectrum is noisy due to fact that the inverse tangents are computed from the *ratio* of imaginary part to real part of the FFT result. Even a small floating rounding off error will amplify the result and manifest incorrectly as useful phase information [4]. To understand, print the first few samples from the FFT result and observe that they are not absolute zeros (they are very small numbers in the order 10^{-16}. Computing inverse tangent will result in incorrect results.

```
>> X(1:5)
ans =
1.0e-16 *
-0.7286 -0.3637 -0.2501i -0.4809 -0.1579i -0.3602 -0.5579i 0.0261 -0.495i
>> atan2(imag(X(1:5)),real(X(1:5)))
ans =
3.1416  -2.5391  -2.8244  -2.1441  -1.5181
```

The solution is to define a tolerance threshold and ignore all the computed phase values that are below the threshold.

Program 1.16: Extracting phase information from FFT - using a tolerance threshold

```
1  X2=X;%store the FFT results in another array
2  %detect noise (very small numbers (eps)) and ignore them
3  threshold = max(abs(X))/10000; %tolerance threshold
4  X2(abs(X)<threshold)=0;%maskout values below the threshold
5  phase=atan2(imag(X2),real(X2))*180/pi; %phase information
6  stem(f,phase); %phase vs frequencies
```

The recomputed phase spectrum is plotted in Figure 1.15. The phase spectrum has correctly registered the $30°$ phase shift at the frequency $f = 10Hz$. The phase spectrum is anti-symmetric ($\phi = -30°$ at $f = -10Hz$), which is expected for real-valued signals.

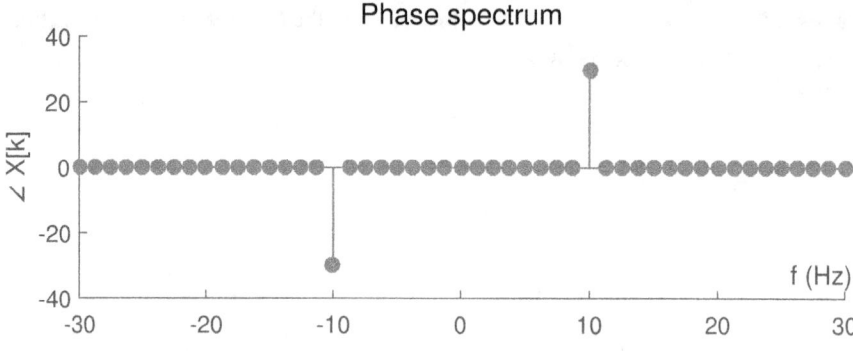

Fig. 1.15: Extracted phase information from the FFT result - recomputed phase spectrum

1.3.3 Reconstructing the time domain signal from the frequency domain samples

Reconstruction of the time domain signal from the frequency domain sample is pretty straightforward. The reconstructed signal, shown in Figure 1.16, has preserved the same initial phase shift and the frequency of the original signal. Note: The length of the reconstructed signal is only 256 sample long (0.8 seconds duration), this is because the size of FFT is considered as $N = 256$. Since the signal is periodic it is not a concern. For more complicated signals, appropriate FFT length (better to use a value that is larger than the length of the signal) need to be used.

Program 1.17: Reconstructing a signal from frequency domain samples

```
1  x_recon = N*ifft(ifftshift(X),N); %reconstructed signal
2  t = [0:1:length(x_recon)-1]/fs; %recompute time index
3  plot(t,x_recon);%reconstructed signal
```

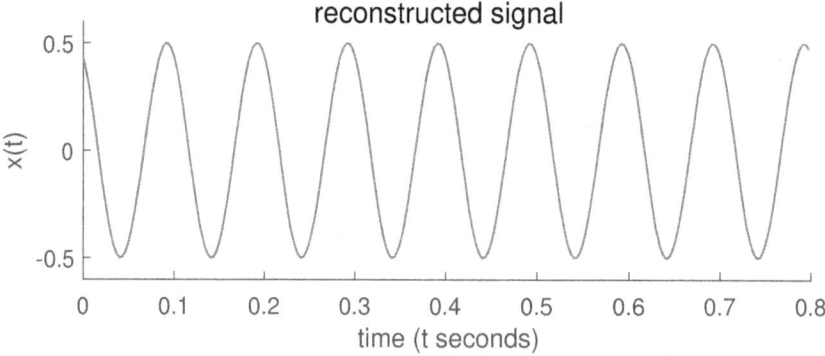

Fig. 1.16: Reconstructed time-domain signal from frequency domain samples

1.4 Power spectral density

Power spectral density (PSD) measures how the power of a signal is distributed over frequency. It is a measure of a signal's power intensity in the frequency domain. The PSD of a signal can be computed as square of the magnitude of the DFT computed in equation 1.27. However, if the signal is stochastic in nature, trying to calculate the spectral components of the signal using DFT will not be valid because, for every realization of the random process, the computed DFT will differ.

For a stochastic signal, if sufficient and accurate statistical model is available, then one can calculate its power spectral density using the Wiener-Khinchin theorem. For a wide-sense stationary random process, the PSD can be calculated as the Fourier transform of the auto-correlation function of the signal.

$$S_{xx}(f) = \mathbb{F}\left[R_{xx}(\tau)\right] = \int_{-\infty}^{\infty} R_{xx}(\tau)e^{-j2\pi f\tau}d\tau \tag{1.29}$$

where, $R_{xx}(\tau)$ is the auto-correlation function of the random process $x(t)$ given by:

$$R_{xx}(\tau) = \mathbb{E}\left(X(t)X(t-\tau)\right) = \int_{-\infty}^{\infty} x(t)x(t+\tau)dt \tag{1.30}$$

We can prove that computing PSD is equivalent to Fourier transform of the auto-correlation function of the stochastic signal $x(t)$ as follows:

$$\mathbb{F}\left[R_{xx}(\tau)\right] = \int_{-\infty}^{\infty} R_{xx}(\tau)e^{-j2\pi f\tau}d\tau \tag{1.31}$$

$$= \int_{-\infty}^{\infty}\int_{-\infty}^{\infty} x(t)x(t+\tau)e^{-j2\pi f\tau}dtd\tau$$

$$= \int_{-\infty}^{\infty} x(t)\underbrace{\int_{-\infty}^{\infty} x(t+\tau)e^{-j2\pi f\tau}d\tau}_{\mathbb{F}[x(t+\tau)]=X(f)e^{j2\pi ft}} dt$$

$$= X(f)\int_{-\infty}^{\infty} x(t)e^{j2\pi ft}dt$$

$$= X(f)X^*(f) = |X(f)|^2$$

For a stochastic signal in analysis, if sufficient statistical details are not available which is often the practical case, one must *estimate* the power spectral density. Several methods are available using which the PSD can be estimated and the Welch method is one among them. Welch method estimates the power spectrum of a signal by dividing it into overlapping blocks, computing the *periodogram* which is the Fourier transform of the estimate of the autocorrelation sequence and averaging the results. The description of the Welch method and its implementation is beyond the scope of this book. The Welch PSD estimate is available in Matlab as the inbuilt function pwelch.

The following function readily plots the Welch spectrum estimate using the pwelch command. The function uses the recommended settings from [5], where the pwelch command is configured with Hanning window, without overlap and an averaging factor of 16. The given function is utilized in chapter chapter 2 for plotting the PSD estimates of signals modulated using various digital modulation techniques, namely, BPSK, QPSK, MSK and GMSK (refer sections 2.10 and 2.11.3)

Program 1.18: *plotWelchPSD.m*: Compute and plot Welch PSD estimates of a stochastic signal

```
function [Pss,f]=plotWelchPSD(SIGNAL,Fs,Fc,COLOR)
%Plot PSD of a carrier modulated SIGNAL using Welch estimate
% SIGNAL - signal vector for which the PSD is plotted
% Fs - Sampling Frequency
% Fc - Center-carrier frequency of the SIGNAL
% COLOR - color character for the plot
ns = max(size(SIGNAL));
na = 16;%averaging factor to plot averaged welch spectrum
w = hanning(floor(ns/na));%Hanning window
%Welch PSD estimate with Hanning window and no overlap
[Pss,f]=pwelch(SIGNAL,w,0,[],Fs,'twosided');
indices = find(f>=Fc & f<4*Fc); %To plot PSD from Fc to 4*Fc
Pss=Pss(indices)/Pss(indices(1)); %normalized psd w.r.t Fc
plot(f(indices)-Fc,10*log10(Pss),COLOR);%normalize frequency axis
```

1.5 Power and energy of a signal

1.5.1 Energy of a signal

In signal processing, a signal is viewed as a function of time. The term *size of a signal* is used to represent *strength of the signal*. It is crucial to know the *size* of a signal used in a certain application. For example, we may be interested to know the amount of electricity needed to power a LCD monitor as opposed to a CRT monitor. Both of these applications are different and have different tolerances. Thus the amount of electricity driving these devices will also be different.

A given signal's size can be measured in many ways. Given a mathematical function (or a signal equivalently), it seems that the area under the curve, described by the mathematical function, is a good measure of describing the size of a signal. A signal can have both positive and negative values. This may render areas that are negative. Due to this effect, it is possible that the computed values cancel each other totally or partially, rendering incorrect result. Thus the metric function of "area under the curve" is not suitable for defining the "size" of a signal. Now, we are left with two options : either 1) computation of the area under the absolute value of the function or 2) computation of the area under the square of the function. The second choice is favored due to its mathematical tractability and its similarity to Euclidean Norm which is used in signal detection techniques (Note: Euclidean norm - otherwise called *L2 norm* or *2-norm* [6] - is often considered in signal detection techniques - on the assumption that it provides a reasonable measure of distance between two points on signal space. It is computed as Euclidean distance in detection theory - see section 3.5.4).

Going by the second choice of viewing the size as the computation of the area under the square of the function, the energy of a continuous-time complex signal $x(t)$ is defined as

$$E_x = \int_{-\infty}^{\infty} |x(t)|^2 dt \tag{1.32}$$

If the signal $x(t)$ is real, the modulus operator in the above equation does not matter. This is called *energy* in signal processing terms. This is also a measure of signal strength. This definition can be applied to any signal (or a vector) irrespective of whether it possesses actual energy (a basic quantitative property as described by physics) or not. If the signal is associated with some physical energy, then the above definition gives the energy content in the signal. If the signal is an electrical signal, then the above definition gives the total energy of the signal (in Joules) dissipated over a 1 Ω resistor.

Actual Energy - the physical quantity

To know the actual energy of the signal E, one has to know the value of load Z the signal is driving and also the nature the electrical signal (voltage or current). For a voltage signal, the above equation has to be scaled by a factor of $1/Z$.

$$E = \frac{E_x}{Z} = \frac{1}{Z} \int_{-\infty}^{\infty} |x(t)|^2 \, dt \tag{1.33}$$

For current signal, it has to be scaled by Z.

$$E = ZE_x = Z \int_{-\infty}^{\infty} |x(t)|^2 \, dt \tag{1.34}$$

where, Z is the impedance driven by the signal $x(t)$, E_x is the signal energy (signal processing term) and E is the Energy of the signal (physical quantity) driving the load Z

Energy in discrete domain

In the discrete domain, the energy of the signal is given by

$$E_x = \sum_{n=-\infty}^{\infty} |x(n)|^2 \tag{1.35}$$

The energy is finite only if the above sum converges to a finite value. This implies that the signal is *squarely-summable*. Such a signal is called finite energy signal. If the given signal does not decay with respect to time (example: a continuous sine wave repeating its cycle infinitely), the energy will be infinite and such a signal is *not squarely-summable* in other words. We need another measurable quantity to circumvent this problem. This leads us to the notion of *Power*

1.5.2 Power of a signal

Power is defined as the amount of energy consumed per unit time. This quantity is useful if the energy of the signal goes to infinity or the signal is *not-squarely-summable*. For non-squarely-summable signals, the power calculated by taking the snapshot of the signal over a specific interval of time as follows

1. Take a snapshot of the signal over some finite time duration
2. Compute the energy of the signal E_x
3. Divide the energy by number of samples taken for computation N
4. Extend the limit of number of samples to infinity $N \to \infty$. This gives the total power of the signal.

In discrete domain, the total power of the signal is given by

$$P_x = \lim_{N \to \infty} \frac{1}{2N+1} \sum_{n=-N}^{n=+N} |x(n)|^2 \tag{1.36}$$

Following equations are different forms of the same computation found in many text books. The only difference is the number of samples taken for computation. The denominator changes according to the number of samples taken for computation.

$$P_x = \lim_{N \to \infty} \frac{1}{2N} \sum_{n=-N}^{n=N-1} |x(n)|^2$$

$$P_x = \lim_{N \to \infty} \frac{1}{N} \sum_{n=0}^{n=N-1} |x(n)|^2$$

$$P_x = \lim_{N \to \infty} \frac{1}{N_1 - N_0 + 1} \sum_{n=N_0}^{n=N_1} |x(n)|^2 \tag{1.37}$$

1.5.3 Classification of signals

A signal can be classified based on its power or energy content. Signals having finite energy are energy signals. Power signals have finite and non-zero power.

Energy signal

A finite energy signal will have zero *total* power. When the energy is finite, the total power will be zero. For example, in equation 1.36, when the limit $N \to \infty$, the energy dilutes to zero over the infinite duration and hence the total power becomes zero.

Power signal

Signals whose total power is finite and non-zero. The energy of the power signal will be infinite. Example: Periodic sequences like sinusoid. A sinusoidal signal has finite, non-zero power but infinite energy. A signal cannot be both an energy signal and a power signal.

Neither an energy signal nor a power signal

Signals can also be a cat on the wall - neither an energy signal nor a power signal. Consider a signal of increasing amplitude defined by $x(n) = n$. For such a signal, both the energy and power will be infinite. Thus, it cannot be classified either as an energy signal or as a power signal.

1.5.4 Computation of power of a signal - simulation and verification

The total power of a signal can be computed using the following equation. For other forms of equations for computing power, refer equation 1.37.

$$P_x = \lim_{N \to \infty} \frac{1}{N} \sum_{n=0}^{n=N-1} |x(n)|^2 \tag{1.38}$$

As a case study, a sine wave of amplitude A and frequency f_c is considered here.

$$x(t) = A \, sin(2\pi f_c t) \tag{1.39}$$

When represented in frequency domain, it will look like the one on the right side plot in the Figure 1.17(a). This is evident from the fact that the sine wave can be mathematically represented by applying Euler's formula.

$$A \sin(2\pi f_c t) = A \frac{e^{j2\pi f_c t} - e^{-j2\pi f_c t}}{2j} \tag{1.40}$$

Taking the Fourier transform of $x(t)$ to represent it in frequency domain,

$$X(f) = F\left[A\sin(2\pi f_c t)\right] = \int_{-\infty}^{\infty} \left[\frac{e^{j2\pi f_c t} - e^{-j2\pi f_c t}}{2j}\right] e^{-j2\pi f t} dt = \frac{A}{2j}\left[\delta(f - f_c) - \delta(f + f_c)\right] \tag{1.41}$$

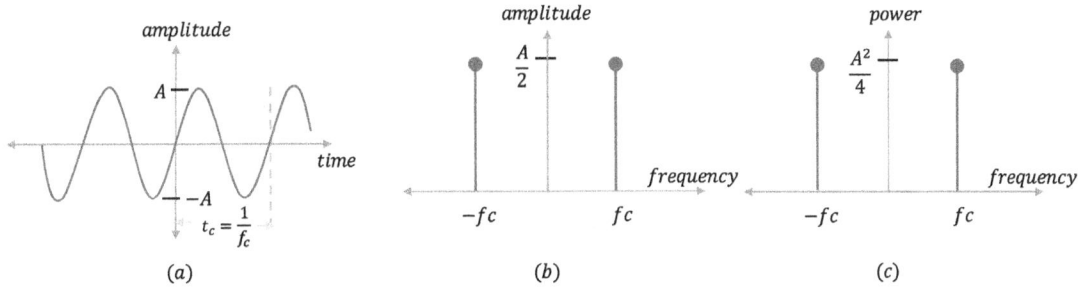

Fig. 1.17: A sinusoid represented in : (a) time domain, (b) frequency domain and (c) its power spectrum

When considering the amplitude part, the above decomposition gives two spikes of amplitude $A/2$ on either side of the frequency domain at f_c and $-f_c$ and this is shown in Figure 1.17(b). Squaring the amplitudes gives the magnitude of power of the individual frequency components. The power spectrum is shown in Figure 1.17(c).

Thus if the pure sine wave is of amplitude $A = 1 \ V$ and frequency $f_c = 100 \ Hz$, the power spectrum will have two spikes of value $A^2/4 = 0.25$ at $100 \ Hz$ and $-100 \ Hz$ frequencies. The total power will be $A^2/4 + A^2/4 = 0.25 + 0.25 = 0.5 \ W$. In order to verify this through simulation, a sine wave of $100 \ Hz$ frequency and amplitude $1 \ V$ is taken for the experiment. The generated sine wave of 3 cycles is plotted in Figure 1.18(a).

Program 1.19: Generating a sinusoid and plotting its power spectrum

```
A=1; %Amplitude of sine wave
fc=100; %Frequency of sine wave
fs=3000; %Sampling frequency - oversampled by the rate of 30
nCyl=3; %Number of cycles of the sinewave

t=0:1/fs:nCyl/fc-1/fs; %Time base
x=-A*sin(2*pi*fc*t); %Sinusoidal function
subplot(1,2,1); plot(t,x);title('Sinusoid of frequency f_c=100 Hz');
xlabel('Time(s)'); ylabel('Amplitude');
```

Fig. 1.18: (a) 10 cycles of the generated sine wave , (b) power spectrum

Matlab's Norm function

Matlab's basic installation comes with norm function. The p-norm in Matlab is computed as

$$NORM(v,p) = \left(\sum_{n=0}^{n=N-1} |x(v)|^p \right)^{1/p} \tag{1.42}$$

By default, the single argument norm function computed 2-norm given as

$$NORM(v) = NORM(v,2) = \left(\sum_{n=0}^{n=N-1} |x(v)|^2 \right)^{1/2} \tag{1.43}$$

To compute the total power of the signal $x[n]$ (as given in equation 1.38), all we have to do is - compute norm(x), square it and divide by the length of the signal. It can be verified that the snippet of code given here, gives the total power as 0.5.

Program 1.20: Computing total power of the generated sinusoid signal from time domain

```
1   L=length(x);
2   P=(norm(x)^2)/L;
3   disp(['Power of the Signal from Time domain ',num2str(P)]);
```

Verifying the total power in frequency domain

Here, the total power is verified by applying *Discrete Fourier Transform (DFT)* on the sinusoidal sequence. The sinusoidal sequence $x[n]$ is represented in frequency domain $X[f]$ using Matlab's FFT function. The power associated with each frequency point is computed as

$$P_x[f] = X[f]X^*[f] \tag{1.44}$$

Finally, the total power is calculated as the sum of all the points in the frequency domain representation. The code will result in the calculated total power equaling a value of 0.5 and the power spectrum plot as given in Figure 1.18(b)

Program 1.21: Computing total power of the generated sinusoid signal from frequency domain

```
1  L=length(x); NFFT=L;
2  X=fftshift(fft(x,NFFT));
3  Px=X.*conj(X)/(L^2); %Power of each freq components
4  fVals=fs*(-NFFT/2:NFFT/2-1)/NFFT;
5  subplot(1,2,2); stem(fVals,Px,'b'); title('Power Spectral Density');
6  xlabel('Frequency (Hz)');ylabel('Power');
```

1.6 Polynomials, convolution and Toeplitz matrices

Convolution operation is ubiquitous in signal processing applications. The mathematics of convolution is strongly rooted in operation on polynomials. The intent of this text is to enhance the understanding on mathematical details of convolution.

1.6.1 Polynomial functions

Polynomial functions are expressions consisting of sum of terms, where each term includes one or more variables raised to a non-negative power and each term may be scaled by a coefficient. Addition, Subtraction and multiplication of polynomials are possible.

Polynomial functions can involve one or more variables. For example, following polynomial expression is a function of variable x. It involves sum of 3 terms where each term is scaled by a coefficient.

$$f(x) = x^2 + 2x + 1 \tag{1.45}$$

Polynomial expression involving two variables x and y is given next.

$$f(x,y) = 2x^4 - 4x^2y + 3xy^2 + 8y^2 + 5 \tag{1.46}$$

1.6.2 Representing single variable polynomial functions

Polynomial functions involving single variable is of specific interest here. In general, a single variable (say x) polynomial is expressed in the following sum of terms form, where $a_0, a_1, a_2, ..., a_{n-1}$ are coefficients of the polynomial.

$$f(x) = a_0 + a_1x + a_2x^2 + a_3x^3 + ... + a_{n-1}x^{n-1} \tag{1.47}$$

The degree or order of the polynomial function is the highest power of x with a non-zero coefficient. The above equation can be written as

$$f(x) = \sum_{i=0}^{n-1} a_i x^i \tag{1.48}$$

It can be represented by a vector of coefficients as $a = [a_0, a_1, a_2, ..., a_{n-1}]$. Polynomials can also be represented using their roots which is a product of linear terms form, as explained next.

1.6.3 Multiplication of polynomials and linear convolution

Mathematical operations like additions, subtractions and multiplications can be performed on polynomial functions. Addition or subtraction of polynomials is straight forward. Multiplication of polynomials is of specific interest in the context of subject discussed here. Let's represent two polynomials represented by the coefficient vectors $a = [a_0, a_1, a_2, ..., a_n]$ and $b = [b_0, b_1, b_2, ..., b_n]$.

$$p(x) = a_0 + a_1 x + a_2 x^2 + a_3 x^3 + ... + a_n x^n \qquad (1.49)$$

$$q(x) = b_0 + b_1 x + b_2 x^2 + a_3 x^3 + ... + b_m x^m \qquad (1.50)$$

The product vector is expressed as

$$p(x).q(x) = a_0 b_0 + (a_1 b_0 + b_1 a_0)x + ... + a_n b_m x^{n+m} \qquad (1.51)$$

or equivalently,

$$p(x).q(x) = (p*q)(x) = a.b = [c_0, c_1, c_2, ..., c_{n+m}] \qquad (1.52)$$

where,

$$c_k = \sum_{i,j:i+j=k} a_i b_j \quad k = 0, 1, ..., n+m \qquad (1.53)$$

Since the subscripts obey the equality $i + j = k$, changing the subscript j to $k - i$ gives

$$c_k = \sum_{i=-\infty}^{\infty} a_i b_{k-i} \quad k = 0, 1, ..., n+m \qquad (1.54)$$

which, when written in terms of array index, provides the most widely used form seen in signal processing text books.

$$c[k] = \sum_{i=-\infty}^{\infty} a[i]b[k-i] \quad k = 0, 1, ..., n+m \qquad (1.55)$$

This operation is referred as *linear convolution*, denoted by the symbol $*$. It is very closely related to other operations on vectors like cross-correlation, auto-correlation and moving average computation. Thus, when we are computing convolution, we are actually multiplying two polynomials. Note, that if the polynomials have N and M terms, their multiplication produces $N + M - 1$ terms.

1.6.4 Toeplitz matrix and convolution

Convolution operation of two sequences can be viewed as multiplying two matrices as explained next. Given an *linear time invariant* (LTI) system with impulse response $h[n]$ and an input sequence $x[n]$, the output of the system $y[n]$ is obtained by *linearly* convolving the input sequence and impulse response.

$$y[k] = h[n] * x[n] = \sum_{i=-\infty}^{\infty} x[i]h[k-i] \quad k = 0, 1, ..., N+M-1 \qquad (1.56)$$

where, the sequence $x[n]$ is of length N and $h[n]$ is of length M. Assume that the sequence $h[n]$ is of length 4 given by $h[n] = [h_0, h_1, h_2, h_3]$ and the sequence $x[n]$ is of length 3 given by $x[n] = [x_0, x_1, x_2]$. The convolution $h[n] * x[n]$ is given by

$$y[k] = h[n] * x[n] = \sum_{i=-\infty}^{\infty} x[i]h[k-i] \quad k = 0, 1, ..., 6 \tag{1.57}$$

Computing each sample in the convolution,

$$y[0] = \sum_{i=-\infty}^{\infty} x[i]h[-i] = x[0]h[0] + 0 + 0$$

$$y[1] = \sum_{i=-\infty}^{\infty} x[i]h[1-i] = x[0]h[1] + x[1]h[0] + 0$$

$$y[2] = \sum_{i=-\infty}^{\infty} x[i]h[2-i] = x[0]h[2] + x[1]h[1] + x[2]h[0]$$

$$y[3] = \sum_{i=-\infty}^{\infty} x[i]h[3-i] = x[0]h[3] + x[1]h[2] + x[2]h[1]$$

$$y[4] = \sum_{i=-\infty}^{\infty} x[i]h[4-i] = x[1]h[3] + x[2]h[1] + 0$$

$$y[5] = \sum_{i=-\infty}^{\infty} x[i]h[5-i] = x[2]h[3] + 0 + 0 \tag{1.58}$$

Note the above result. See how the series $x[n]$ and $h[n]$ multiply with each other from the opposite directions. This gives the reason on why we have to reverse one of the sequences and shift one step a time to do the convolution operation.

Thus, graphically, in convolution, one of the sequences (say $h[n]$) is reversed (reflected along the y axis). It is delayed to the extreme left where there are no overlaps between the two sequences. Now, the sample offset of $h[n]$ is increased 1 step at a time. At each step, the overlapping portions of $h[n]$ and $x[n]$ are multiplied and summed. This process is repeated until the sequence $h[n]$ is slid to the extreme right where no more overlaps between $h[n]$ and $x[n]$ are possible.

Representing the equation 1.58 in matrix form,

$$\begin{bmatrix} y[0] \\ y[1] \\ y[2] \\ y[3] \\ y[4] \\ y[5] \end{bmatrix} = \begin{bmatrix} h[0] & 0 & 0 \\ h[1] & h[0] & 0 \\ h[2] & h[1] & h[0] \\ h[3] & h[2] & h[1] \\ 0 & h[3] & h[2] \\ 0 & 0 & h[3] \end{bmatrix} \begin{bmatrix} x[0] \\ x[1] \\ x[2] \end{bmatrix} \tag{1.59}$$

When the sequences $h[n]$ and $x[n]$ are represented as matrices, the convolution operation can be equivalently represented as

$$y = h * x = x * h = \begin{bmatrix} h_0 & 0 & 0 \\ h_1 & h_0 & 0 \\ h_2 & h_1 & h_0 \\ h_3 & h_2 & h_1 \\ 0 & h_3 & h_2 \\ 0 & 0 & h_3 \end{bmatrix} \begin{bmatrix} x_0 \\ x_1 \\ x_2 \end{bmatrix} \tag{1.60}$$

The matrix representing the incremental delays of $h[n]$ used in the above equation is a special form of matrix called *Toeplitz matrix*. Toeplitz matrices have constant entries along their diagonals. Toeplitz matrices are used to model systems that possess shift invariant properties. The property of shift invariance is evident from the matrix structure itself. Since we are modeling an LTI system [7], Toeplitz matrices are our natural choice. On a side note, a special form of Toeplitz matrix called *Circulant Matrix* is used in applications involving *circular convolution* and *Discrete Fourier Transform (DFT)* [8].

Representing the construction of Toeplitz matrix, in equation 1.60, as a function $T(h)$,

$$T(h) = \begin{bmatrix} h_0 & 0 & 0 \\ h_1 & h_0 & 0 \\ h_2 & h_1 & h_0 \\ h_3 & h_2 & h_1 \\ 0 & h_3 & h_2 \\ 0 & 0 & h_3 \end{bmatrix} \tag{1.61}$$

the convolution of h and x is simply a matrix multiplication of Toeplitz matrix $T(h)$ and the matrix representation of x denoted as X

$$y = h * x = x * h = \begin{bmatrix} h_0 & 0 & 0 \\ h_1 & h_0 & 0 \\ h_2 & h_1 & h_0 \\ h_3 & h_2 & h_1 \\ 0 & h_3 & h_2 \\ 0 & 0 & h_3 \end{bmatrix} \begin{bmatrix} x_0 \\ x_1 \\ x_2 \end{bmatrix} = T(h).X \tag{1.62}$$

One can quickly vectorize the convolution operation in matlab by using Toeplitz matrices [9] as shown here.

```
y=toeplitz([h0 h1 h2 h3 0 0],[h0 0 0])*x.'
```

1.7 Methods to compute convolution

Mathematical details of convolution, its relationship to polynomial multiplication and the application of Toeplitz matrices in computing linear convolution are discussed in the previous section (1.6) of this chapter. A short survey of different techniques to compute discrete linear convolution is given here.

Given a linear time invariant (LTI) system with impulse response $h[n]$ and an input sequence $x[n]$, the output of the system $y[n]$ is obtained by convolving the input sequence and impulse response.

$$y[k] = h[n] * x[n] = \sum_{i=-\infty}^{\infty} x[i]h[k-i] \quad k = 0, 1, ..., N+M \tag{1.63}$$

where, the sequence $x[n]$ is of length N and $h[n]$ is of length M. Matlab contains an in-built function called conv.m for computing convolution. If the signal processing toolbox is unavailable, the following methods can be substituted for computing convolution between any two given sequences.

1.7.1 Method 1: Brute-force method

As shown in section 1.6.3, the convolution equation 1.63 can simply be interpreted as polynomial multiplication and hence it can be implemented using nested for-loops. However this method consumes the highest computational time of all the methods given here. Typically, the computational complexity is $O(n^2)$ time.

Program 1.22: *conv_brute_force.m*: Brute force method to compute convolution

```
function y = conv_brute_force(x,h)
%Brute force method to compute convolution

N=length(x); M=length(h);
y = zeros(1,N+M-1);
for i = 1:N,
    for j = 1:M,
        y(i+j-1) = y(i+j-1) + x(i) * h(j);
    end
end
end
```

1.7.2 Method 2: Using Toeplitz matrix

When the sequences $h[n]$ and $x[n]$ are represented as matrices, the convolution operation can be equivalently represented as

$$y = h * x = x * h = Toeplitz(h).X = \mathbf{T}(h).\mathbf{X} \tag{1.64}$$

When the convolution of two sequences of lengths N and p is computed, the Toeplitz matrix $\mathbf{T}(h)$ is of length $(N+p-1) \times (p)$.

The following generic function constructs a Toeplitz matrix of length $(N+p-1) \times (p)$ from a given sequence of length N. Refer section 1.6.4 for the mathematical details of Toeplitz matrix and its relationship to convolution operation.

Program 1.23: *convMatrix.m*: Function to construct a Toeplitz matrix of length $(N+p-1) \times (p)$

```
function [H]=convMatrix(h,p)
%Construct the convolution matrix of size (N+p-1)x p from the input
%matrix h of size N.
  h=h(:).';
  col=[h zeros(1,p-1)]; row=[h(1) zeros(1,p-1)];
  H=toeplitz(col,row);
end
```

A generalized function called `convolve`, given here, finds the convolution of two sequences of arbitrary lengths. It makes uses of the function `convMatrix` that was just described above.

Program 1.24: *convolve.m*: Function to compute convolution of two sequences

```
function [y]=convolve(h,x)
%Convolve two sequences h and x of arbitrary lengths: y=h*x
    H=convMatrix(h,length(x)); %see convMatrix.m
    y=H*conj(x'); %equivalent to conv(h,x) inbuilt function
end
```

1.7.3 Method 3: Using FFT to compute convolution

Computation of convolution using FFT (Fast Fourier Transform) has the advantage of reduced computational complexity when the length of the input vectors are large. To compute convolution, take FFT of the two sequences x and h with FFT length set to convolution output length $length(x) + length(h) - 1$, multiply the results and convert back to time-domain using IFFT (Inverse Fast Fourier Transform). Note that FFT is a direct implementation of *circular convolution* in time domain. Here ,we are attempting to compute linear convolution using circular convolution (or FFT) with zero-padding either one of the input sequence. This causes inefficiency when compared to circular convolution. Nevertheless, this method still provides $O\left(\frac{N}{log_2 N}\right)$ savings over brute-force method.

$$y(n) = IFFT[FFT_L(\mathbf{x}) * FFT_L(\mathbf{h})] \qquad 2^L \geq |N + M - 1| \tag{1.65}$$

Usually, the following algorithm is suffice which ignores additional zeros in the output terms.

$$y(n) = IFFT[FFT_L(\mathbf{X}) * FFT_L(\mathbf{H})] \qquad L = N + M - 1 \tag{1.66}$$

Matlab code snippet for implementing the above algorithm is given next.

```
y=ifft(fft(x,L).*(fft(h,L))) %convolution using FFT and IFFT
```

Test and comparison

Let's test the convolution methods, especially the method 2 (convolution using Toeplitz matrix transformation) and method 3 (convolution using FFT) by comparing them against Matlab's standard conv function.

Program 1.25: Comparing different methods for computing convolution

```
1  x=randn(1,7)+1i*randn(1,7) %Create random vectors for test
2  h=randn(1,3)+1i*randn(1,3) %Create random vectors for test
3  L=length(x)+length(h)-1; %length of convolution output
4
5  y1=convolve(h,x) %Convolution Using Toeplitz matrix
6  y2=ifft(fft(x,L).*(fft(h,L))).' %Convolution using FFT
7  y3=conv(h,x) %Matlab's standard function
```

On comparing method 2 (output $y1$) and method 3 (output $y2$) with Matlab's standard convolution function (output $y3$), it is found that all three methods yield identical results. Results obtained on a sample run are given here.

$$x = \begin{bmatrix} 0.5404 + 0.2147i \\ -0.0915 + 2.0108i \\ -0.7603 + 0.0256i \\ -0.6936 + 0.3083i \\ 1.2815 - 0.9382i \\ -0.8097 + 1.6742i \\ -1.2368 + 0.1250i \end{bmatrix} ; \ h = \begin{bmatrix} 0.5301 + 0.3891i \\ -0.9521 - 1.1560i \\ 0.8540 + 0.0397i \end{bmatrix} ; \ y1 = y2 = y3 = \begin{bmatrix} 0.2029 + 0.3241i \\ -1.0973 + 0.2012i \\ 2.4516 - 1.8860i \\ 0.1076 + 2.4617i \\ 1.4109 + 0.5012i \\ -3.9900 + 0.2200i \\ 3.1337 - 1.8233i \\ 0.5639 + 2.7084i \\ -1.0613 + 0.0576i \end{bmatrix} \tag{1.67}$$

1.7.4 Miscellaneous methods

If the input sequence is of infinite length or very large as in many real time applications, block processing methods like *Overlap-Add* and *Overlap-Save* can be used to compute convolution in a faster and efficient way. There exist standard algorithms for computing convolution that greatly reduce the computational complexity. Some of them are given here. Refer [10] for more details.

- Cook-Toom Algorithm
- Modified Cook-Toom Algorithm
- Winograd Algorithm
- Modified Winograd Algorithm
- Iterated Convolution

1.8 Analytic signal and its applications

1.8.1 Analytic signal and Fourier transform

Fourier Transform of a real-valued signal is complex-symmetric. It implies that the content at negative frequencies are redundant with respect to the positive frequencies. In their works, Gabor [11] and Ville [12], aimed to create an *analytic signal* by removing redundant negative frequency content resulting from the Fourier transform. The analytic signal is complex-valued but its spectrum is one-sided (only positive frequencies) that preserves the spectral content of the original real-valued signal. Using an analytic signal instead of the original real-valued signal, has proven to be useful in many signal processing applications. For example, in spectral analysis, use of analytic signal in-lieu of the original real-valued signal mitigates estimation biases and eliminates cross-term artifacts due to negative and positive frequency components [13].

1.8.1.1 Continuous-time analytic signal

Let $x(t)$ be a real-valued non-band-limited finite energy signal, for which we wish to construct a corresponding analytic signal $z(t)$. The *continuous time fourier transform* (CTFT) of $x(t)$ is given by

$$X(f) = \int_{-\infty}^{\infty} x(t)e^{-j2\pi ft}dt \tag{1.68}$$

Let's say the magnitude spectrum of $X(f)$ is as shown in Figure 1.19(a). We note that the signal $x(t)$ is a real-valued and its magnitude spectrum $|X(f)|$ is symmetric and extends infinitely in the frequency domain.

As mentioned before, an analytic signal can be formed by suppressing the negative frequency contents of the Fourier Transform of the real-valued signal. That is, in frequency domain, the spectral content $Z(f)$ of the analytic signal $z(t)$ is given by

$$Z(f) = \begin{cases} X(0) & for\ f = 0 \\ 2X(f) & for\ f > 0 \\ 0 & for\ f < 0 \end{cases} \tag{1.69}$$

The corresponding spectrum of the resulting analytic signal is shown in Figure 1.19(b).

Since the spectrum of the analytic signal is one-sided, the analytic signal will be complex valued in the time domain, hence the analytic signal can be represented in terms of real and imaginary components as $z(t) = z_r(t) + jz_i(t)$. Since the spectral content is preserved in an analytic signal, it turns out that the real part

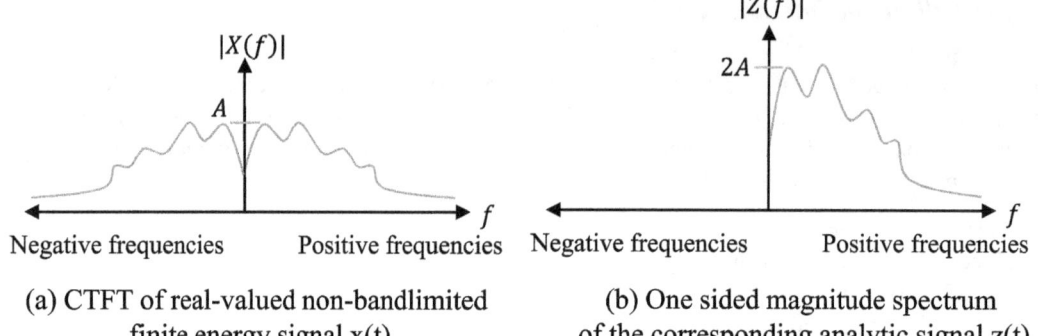

(a) CTFT of real-valued non-bandlimited
finite energy signal x(t)

(b) One sided magnitude spectrum
of the corresponding analytic signal z(t)

Fig. 1.19: (a) Spectrum of continuous signal x(t) and (b) spectrum of analytic signal z(t)

of the analytic signal in time domain is essentially the original real-valued signal itself ($z_r(t) = x(t)$). Then, what takes the place of the imaginary part ? What accompanies $x(t)$, that occupies the imaginary part in the resulting analytic signal ? Summarizing as equation,

$$z(t) = z_r(t) + jz_i(t) \tag{1.70}$$

$$z_r(t) = x(t) \quad z_i(t) = ?? \tag{1.71}$$

It is interesting to note that Hilbert transform [14] can be used to find a companion function (imaginary part in the equation 1.71) to a real-valued signal such that the real signal can be analytically extended from the real axis to the upper half of the complex plane . Denoting Hilbert transform as $HT\{\cdot\}$, the analytic signal is given by

$$z(t) = z_r(t) + jz_i(t) = x(t) + jHT\{x(t)\} \tag{1.72}$$

From these discussions, we can see that an analytic signal $z(t)$ for a real-valued signal $x(t)$, can be constructed using two approaches.

- **Frequency domain approach**: The one-sided spectrum of $z(t)$ is formed from the two-sided spectrum of the real-valued signal $x(t)$ by applying equation 1.69.
- **Time domain approach**: Using Hilbert transform approach given in equation 1.72.

One important property of an analytic signal is that its real and imaginary components are orthogonal.

$$\int_{-\infty}^{\infty} z_i(t)z_r(t) = 0 \tag{1.73}$$

1.8.1.2 Discrete-time analytic signal

Since we are in the digital era, we are more interested in discrete-time signal processing. Consider a continuous real-valued signal $x(t)$, that gets sampled at interval T seconds and results in N real-valued discrete samples $x[n]$, i.e, $x[n] = x(nT)$. The spectrum of the continuous signal is shown in Figure 1.20(a). The spectrum of $x[n]$ that results from the process of periodic sampling is given in Figure 1.20(b). The spectrum of discrete-time signal $x[n]$ can be obtained by *discrete-time fourier transform* (DTFT) .

$$X(f) = T \sum_{n=0}^{N-1} x[n]e^{-j2\pi fnT} \tag{1.74}$$

At this point, we would like to construct a discrete-time analytic signal $z[n]$ from the real-valued sampled signal $x[n]$. We wish the analytic signal is complex valued $z[n] = z_r[n] + jz_i[n]$ and should satisfy the following two desired properties

- The real part of the analytic signal should be same as the original real-valued signal.

$$z_r[n] = x[n] \tag{1.75}$$

- The real and imaginary part of the analytic signal should satisfy the property of orthogonality.

$$\sum_{n=0}^{N-1} z_r[n]z_i[n] = 0 \tag{1.76}$$

In frequency domain approach for the continuous-time case, we saw that an analytic signal is constructed by suppressing the negative frequency components from the spectrum of the real signal. We cannot do this for our periodically sampled signal $x[n]$. Periodic mirroring nature of the spectrum prevents one from suppressing the negative components. If we do so, it will suppress the entire spectrum. One solution to this problem is to set the negative half of each spectral period to zero. The resulting spectrum of the analytic signal is shown in Figure 1.20(c).

Given a record of samples $x[n]$ of even length N, the procedure to construct the analytic signal $z[n]$ is as follows [15].

- Compute the N-point DTFT of $x[n]$ using FFT
- N-point periodic one-sided analytic signal is computed by the following transform

$$Z[m] = \begin{cases} X[0] & for\ m = 0 \\ 2X[m] & for\ 1 \leq m \leq \frac{N}{2} - 1 \\ X[\frac{N}{2}] & for\ m = \frac{N}{2} \\ 0 & for\ \frac{N}{2} + 1 \leq m \leq N - 1 \end{cases} \tag{1.77}$$

- Finally, the analytic signal $z[n]$ is obtained by taking the inverse DTFT of $Z[m]$

$$z[n] = \frac{1}{NT} \sum_{m=0}^{N-1} z[m]\ exp\left(j2\pi mn/N\right) \tag{1.78}$$

This method satisfies both the desired properties listed in equations 1.75 and 1.76. The given procedure can be coded in Matlab using the FFT function. Given a record of N real-valued samples $x[n]$, the corresponding analytic signal $z[n]$ can be constructed as given next.

Program 1.26: *analytic_signal.m*: Generating an analytic signal for a given real-valued signal

```
1  function z = analytic_signal(x)
2  %Generate analytic signal using frequency domain approach
3  x = x(:); %serialize
4  N = length(x);
5  X = fft(x,N);
6  z = ifft([X(1); 2*X(2:N/2); X(N/2+1); zeros(N/2-1,1)],N);
7  end
```

To test this function, we create a 5 seconds record of a real-valued sine signal and pass it as an argument to the function. The resulting analytic signal is constructed and its orthogonal components are plotted in Figure 1.21. From this plot, we can see that the real part of the analytic signal is identical to the original signal and

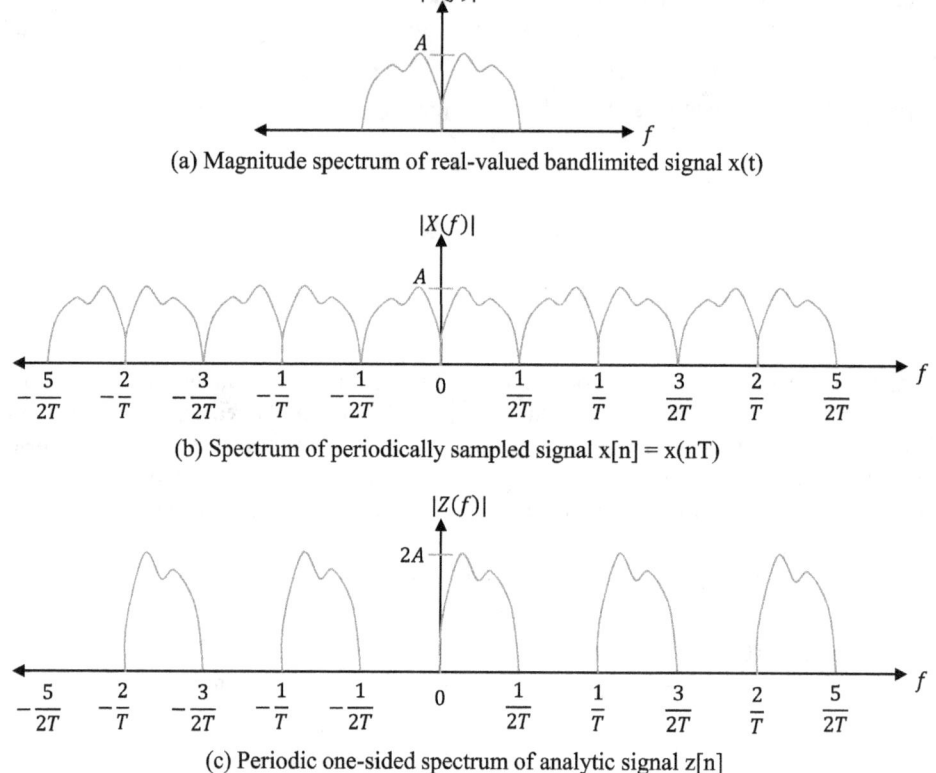

(a) Magnitude spectrum of real-valued bandlimited signal x(t)

(b) Spectrum of periodically sampled signal x[n] = x(nT)

(c) Periodic one-sided spectrum of analytic signal z[n]

Fig. 1.20: (a) CTFT of continuous signal x(t), (b) Spectrum of x[n] resulted due to periodic sampling and (c) Periodic one-sided spectrum of analytical signal z[n].

the imaginary part of the analytic signal is $-90°$ phase shifted version of the original signal. We note that the analytic signal's imaginary part is Hilbert transform of its real part.

Program 1.27: *test_analytic_signal.m*: Check and investigate components of an analytic signal

```
1   %Test routine to check analytic_signal function
2   t=0:0.001:0.5-0.001;
3   x = sin(2*pi*10*t); %real-valued f = 10 Hz
4   subplot(2,1,1); plot(t,x);%plot the original signal
5   title('x[n] - real-valued signal'); xlabel('n'); ylabel('x[n]');
6
7   z = analytic_signal(x); %construct analytic signal
8   subplot(2,1,2); plot(t, real(z), 'k'); hold on;
9   plot(t, imag(z), 'r');
10  title('Components of Analytic signal');
11  xlabel('n'); ylabel('z_r[n] and z_i[n]');
12  legend('Real(z[n])','Imag(z[n])');
```

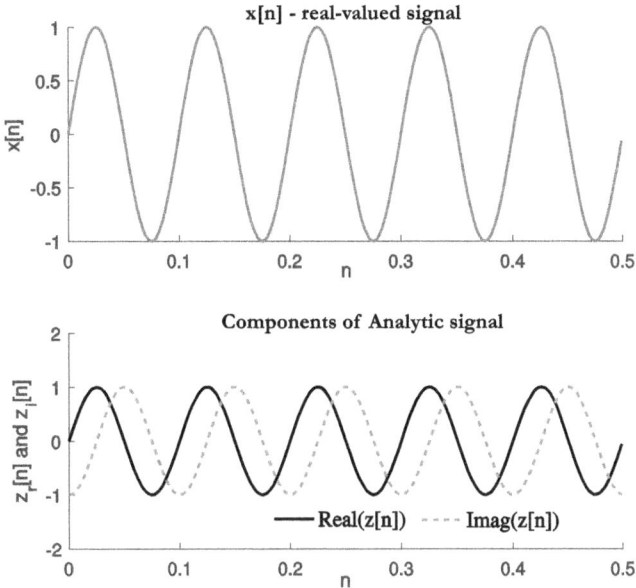

Fig. 1.21: Components of analytic signal for a real-valued sine function

1.8.1.3 Hilbert Transform using Fourier Transform

Matlab's signal processing toolbox has an inbuilt function to compute the analytic signal. The in-built function is called hilbert. We should note that the inbuilt hilbert function in Matlab returns the analytic signal $z[n]$ and not the hilbert transform of the signal $x[n]$. To get the hilbert transform, we should simply get the imaginary part of the analytic signal. Since we have written our own function to compute the analytic signal, getting the hilbert transform of a real-valued signal goes like this.

```
x_hilbert = imag(analytic_signal(x))
```

1.8.2 Applications of analytic signal

Hilbert transform and the concept of analytic signal has several applications. Two of them are detailed next.

1.8.2.1 Extracting instantaneous amplitude, phase, frequency

The concept of instantaneous amplitude/phase/frequency is fundamental to information communication and appears in many signal processing applications. We know that a monochromatic signal of form $x(t) = a\cos(\omega t + \phi)$ cannot carry any information. To carry information, the signal need to be modulated. Take for example the case of amplitude modulation, in which a positive real-valued signal $m(t)$ modulates a carrier $\cos(\omega_c t)$. That is, the amplitude modulation is effected by multiplying the information bearing signal $m(t)$ with the carrier signal $\cos(\omega_c t)$.

$$x(t) = m(t)\cos(\omega_c t) \tag{1.79}$$

Here, ω_c is the *angular frequency* of the signal measured in radians/sec and is related to the *temporal frequency* f_c as $\omega_c = 2\pi f_c$. The term $m(t)$ is also called *instantaneous amplitude*.

Similarly, in the case of phase or frequency modulations, the concept of instantaneous phase or instantaneous frequency is required for describing the modulated signal.

$$x(t) = a\cos\left(\phi(t)\right) \tag{1.80}$$

Here, a is the constant amplitude factor the effects no change in the envelope of the signal and $\phi(t)$ is the instantaneous phase which varies according to the information. The *instantaneous angular frequency* is expressed as the derivative of *instantaneous phase*.

$$\omega(t) = \frac{d}{dt}\phi(t) \tag{1.81}$$

$$f(t) = \frac{1}{2\pi}\frac{d}{dt}\phi(t) \tag{1.82}$$

Generalizing these concepts, if a signal is expressed as

$$x(t) = a(t)\cos\left(\phi(t)\right) \tag{1.83}$$

- The instantaneous amplitude or the *envelope* of the signal is given by $a(t)$.
- The instantaneous phase is given by $\phi(t)$.
- The instantaneous angular frequency is derived as $\omega(t) = \frac{d}{dt}\phi(t)$.
- The instantaneous temporal frequency is derived as $f(t) = \frac{1}{2\pi}\frac{d}{dt}\phi(t)$.

Let's consider the following problem statement: An amplitude modulated signal is formed by multiplying a sinusoidal information and a linear frequency chirp. The information content is expressed as $a(t) = 1 + 0.7\sin(2\pi 3t)$ and the linear frequency chirp is made to vary from $20\,Hz$ to $80\,Hz$. Given the modulated signal, extract the instantaneous amplitude (envelope), instantaneous phase and the instantaneous frequency.

The solution can be as follows. We note that the modulated signal is a real-valued signal. We also take note of the fact that amplitude/phase and frequency can be easily computed if the signal is expressed in complex form. Which transform should we use such that the we can convert a real signal to the complex plane without altering the required properties ? Answer: Apply Hilbert transform and form the analytic signal on the complex plane. Figure 1.22 illustrates this concept.

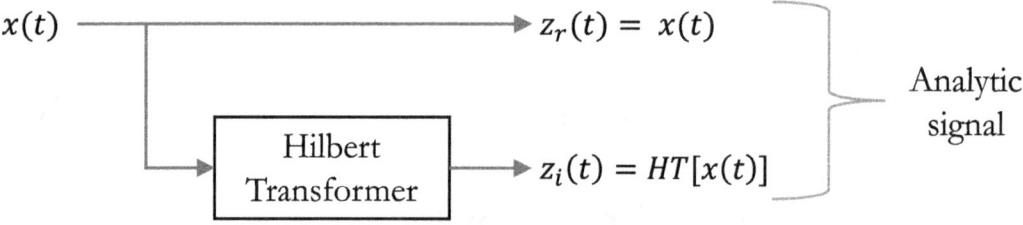

Fig. 1.22: Converting a real-valued signal to complex plane using Hilbert Transform

If we express the real-valued modulated signal $x(t)$ as an analytic signal, it is expressed in complex plane as

$$z(t) = z_r(t) + jz_i(t) = x(t) + jHT\{x(t)\} \tag{1.84}$$

where, $HT\{\cdot\}$ represents the Hilbert Transform operation. Now, the required parameters are very easy to obtain.

- The instantaneous amplitude is computed in the complex plane as

$$a(t) = |z(t)| = \sqrt{z_r^2(t) + z_i^2(t)} \tag{1.85}$$

- The instantaneous phase is computed in the complex plane as

$$\phi(t) = \angle z(t) = arctan\left[\frac{z_i(t)}{z_r(t)}\right] \tag{1.86}$$

- The instantaneous temporal frequency is computed in the complex plane as

$$f(t) = \frac{1}{2\pi}\frac{d}{dt}\phi(t) \tag{1.87}$$

Once we know the instantaneous phase, the carrier can be regenerated as $cos[\phi(t)]$. The regenerated carrier is often referred as *temporal fine structure (TFS)* in acoustic signal processing [16]. The following Matlab code demonstrates the extraction procedure and the resulting plots are shown in Figure 1.23.

Program 1.28: *extract_envelope_phase.m*:Envelope and instantaneous phase extraction from analytic signal

```
1   %Demonstrate extraction of instantaneous amplitude and
2   %phase from the analytic signal constructed from a real-valued
3   %modulated signal.
4   fs = 600; %sampling frequency in Hz
5   t = 0:1/fs:1-1/fs; %time base
6   a_t = 1.0 + 0.7 * sin(2.0*pi*3.0*t) ; %information signal
7   c_t = chirp(t,20,t(end),80); %chirp carrier
8   x = a_t .* c_t; %modulated signal
9
10  subplot(2,1,1); plot(x);hold on; %plot the modulated signal
11
12  z = analytic_signal(x); %form the analytical signal
13  inst_amplitude = abs(z); %envelope extraction
14  inst_phase = unwrap(angle(z));%inst phase
15  inst_freq = diff(inst_phase)/(2*pi)*fs;%inst frequency
16
17  %Regenerate the carrier from the instantaneous phase
18  regenerated_carrier = cos(inst_phase);
19
20  plot(inst_amplitude,'r'); %overlay the extracted envelope
21  title('Modulated signal and extracted envelope');
22  xlabel('n'); ylabel('x(t) and |z(t)|');
23  subplot(2,1,2); plot(cos(inst_phase));
24  title('Extracted carrier or TFS');
25  xlabel('n'); ylabel('cos[\omega(t)]');
```

1.8.2.2 Phase demodulation using Hilbert transform

In communication systems, different types of modulations are available. They are mainly categorized as: amplitude modulation and phase modulation / frequency modulation. In amplitude modulation, the information

Fig. 1.23: Amplitude modulation using a chirp signal and extraction of envelope and TFS

is encoded as variations in the amplitude of a carrier signal. Demodulation of an amplitude modulated signal, involves extraction of the envelope of the modulated signal (see section 1.8.2.1).

In phase modulation, the information is encoded as variations in the phase of the carrier signal. In its generic form, a phase modulated signal expressed as an information-bearing sinusoidal signal modulating another sinusoidal carrier signal is given by

$$x(t) = A cos \left[2\pi f_c t + \beta + \alpha sin \left(2\pi f_m t + \theta \right) \right] \tag{1.88}$$

where, $m(t) = \alpha sin \left(2\pi f_m t + \theta \right)$ represents the information-bearing modulating signal, with the following parameters

- α - amplitude of the modulating sinusoidal signal.
- f_m - frequency of the modulating sinusoidal signal.
- θ - phase offset of the modulating sinusoidal signal.

The carrier signal has the following parameters

- A - amplitude of the carrier.
- f_c - frequency of the carrier and $f_c >> f_m$.
- β - phase offset of the carrier.

The phase modulated signal shown in equation 1.88, can be simply expressed as

$$x(t) = A cos \left[\phi(t) \right] \tag{1.89}$$

Here, $\phi(t)$ is the instantaneous phase that varies according to the information signal $m(t)$. A phase modulated signal of form $x(t)$ can be demodulated by forming an analytic signal by applying hilbert transform and then extracting the instantaneous phase. Extraction of instantaneous phase of a signal was discussed in section 1.8.2.1.

We note that the instantaneous phase is $\phi(t) = 2\pi f_c t + \beta + \alpha sin \left(2\pi f_m t + \theta \right)$ is linear in time, that is proportional to $2\pi f_c t$. This linear offset needs to be subtracted from the instantaneous phase to obtain the

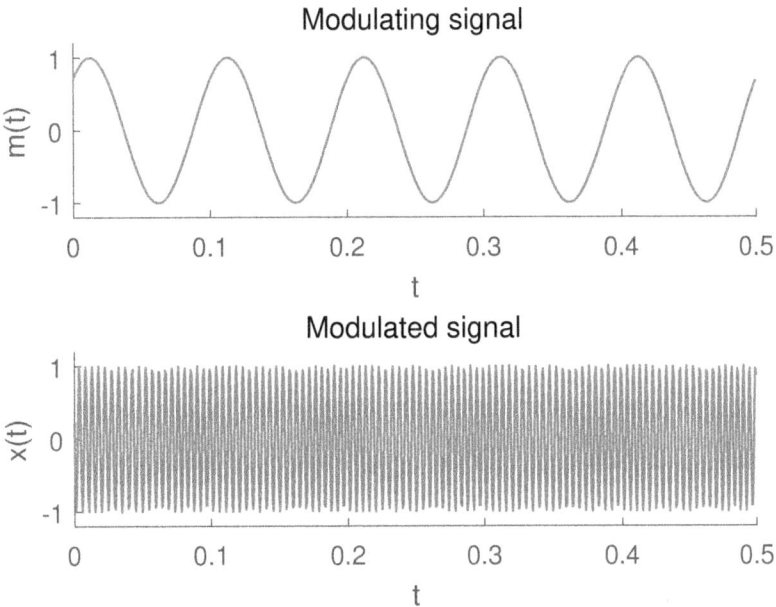

Fig. 1.24: Phase modulation - modulating signal and modulated (transmitted) signal

information bearing modulated signal. If the carrier frequency is known at the receiver, this can be done easily. If not, the carrier frequency term $2\pi f_c t$ needs to be estimated using a linear fit of the unwrapped instantaneous phase.

The following Matlab code demonstrates all these methods. The resulting plots are shown in Figures 1.24 and 1.25.

Program 1.29: *phase_demod_app.m*: Demodulate a phase modulated signal using Hilbert Transform

```
1   %Demonstrate simple Phase Demodulation using Hilbert transform
2   clearvars; clc;
3   fc = 210; %carrier frequency
4   fm = 10; %frequency of modulating signal
5   alpha = 1; %amplitude of modulating signal
6   theta = pi/4; %phase offset of modulating signal
7   beta = pi/5; %constant carrier phase offset
8   receiverKnowsCarrier= 'False';
9   %Set True if receiver knows carrier frequency & phase offset
10
11  fs = 8*fc; %sampling frequency
12  duration = 0.5; %duration of the signal
13  t = 0:1/fs:duration-1/fs; %time base
14
15  %Phase Modulation
16  m_t = alpha*sin(2*pi*fm*t + theta); %modulating signal
17  x = cos(2*pi*fc*t + beta + m_t ); %modulated signal
18
19  figure(); subplot(2,1,1); plot(t,m_t) %plot modulating signal
20  title('Modulating signal'); xlabel('t'); ylabel('m(t)')
```

```
21  subplot(2,1,2); plot(t,x) %plot modulated signal
22  title('Modulated signal'); xlabel('t');ylabel('x(t)')
23
24  %Add AWGN noise to the transmitted signal
25  nMean = 0; nSigma = 0.1; %noise mean and sigma
26  n = nMean + nSigma*randn(size(t)); %awgn noise
27  r = x + n; %noisy received signal
28
29  %Demodulation of the noisy Phase Modulated signal
30  z= hilbert(r); %form the analytical signal from the received vector
31  inst_phase = unwrap(angle(z)); %instaneous phase
32
33  %If receiver knows the carrier freq/phase perfectly
34  if strcmpi(receiverKnowsCarrier,'True')
35      offsetTerm = 2*pi*fc*t+beta;
36  else %else, estimate the subtraction term
37      p = polyfit(t,inst_phase,1);%linearly fit the instaneous phase
38      %re-evaluate the offset term using the fitted values
39      estimated = polyval(p,t); offsetTerm = estimated;
40  end
41  demodulated = inst_phase - offsetTerm;
42  figure(); plot(t,demodulated); %demodulated signal
43  title('Demodulated signal'); xlabel('n'); ylabel('\hat{m(t)}');
```

Fig. 1.25: Demodulated signal from the noisy received-signal and when the carrier is unknown at the receiver

1.9 Choosing a filter : FIR or IIR : understanding the design perspective

Choosing the best filter for implementing a signal processing block is a challenge often faced by communication systems development engineer. There exists two different types of *linear time invariant* (LTI) filters from transfer function standpoint : *finite impulse response* (FIR) and *infinite impulse response* (IIR) filters and myriad design techniques for designing them. The mere fact that there exists so many techniques for designing a filter, suggests that there is no single optimal filter design. One has to weigh-in the pros and cons of choosing a filter design by considering the factors discussed here.

1.9.1 Design specification

A filter design starts with a specification. We may have a simple specification that just calls for removing an unwanted frequency component or it can be a complete design specification that calls for various parameters like - amount of ripples allowed in passband, stop band attenuation, transition width etc. The design specification usually calls for satisfying one or more of the following:

- desired magnitude response - $|H_{spec}(\omega)|$
- desired phase response - $\angle H_{spec}(\omega)$
- tolerance specifications - that specifies how much the filter response is allowed to vary when compared with ideal response. Examples include how much ripples allowed in passband, stop band etc.

Given the specifications above, the goal of a filter design process is to choose the parameters M, N, $\{b_k\}$ and $\{a_k\}$ such that the *transfer function* of the filter

$$H(z) = \frac{\sum_{i=0}^{M} b_k z^{-1}}{1 + \sum_{i=1}^{N} a_k z^{-1}} = \frac{\prod_i (z - z_i)}{\prod_j (z - p_j)} \tag{1.90}$$

yields the desired response: $H(\omega) \approx H_{spec}(\omega)$. In other words, the design process also involves choosing the number and location of the zeros $\{z_i\}$ and poles $\{p_j\}$ in the pole-zero plot. In Matlab, given the transfer function of a filter, the filtering process can be simulated using the `filter` command. The command `y=filter(b,a,x)` filters the input data x using the transfer function defined by the numerator and denominator coefficients $\{b_k\}$ and $\{a_k\}$ respectively.

Two types of filter can manifest from the given transfer function given in equation 1.90.

- When $N = 0$, there is no feedback in the filter structure (Figure 1.26(a)), no poles in the pole-zero plot (in fact all the poles sit at the origin). The impulse response of such filter dies out (becomes zero) beyond certain point of time and it is classified as *finite impulse response (FIR)* filter. It provides *linear phase characteristic in the passband*.
- When $N > 0$, the filter structure is characterized by the presence of feedback elements (Figure 1.26(b)). Due to the presence of feedback elements, the impulse response of the filter may not become zero beyond certain point, but continues indefinitely and hence the name *infinite impulse response* (IIR) filter.
- **Caution:**In most cases, the presence of feedback elements provide infinite impulse response. It is not always true. There are some exceptional cases where the presence of feedback structure may result in finite impulse response. For example, a moving average filter will have a finite impulse response. The output of a moving average filter can be described using a recursive formula, which will result in a structure with feedback elements.

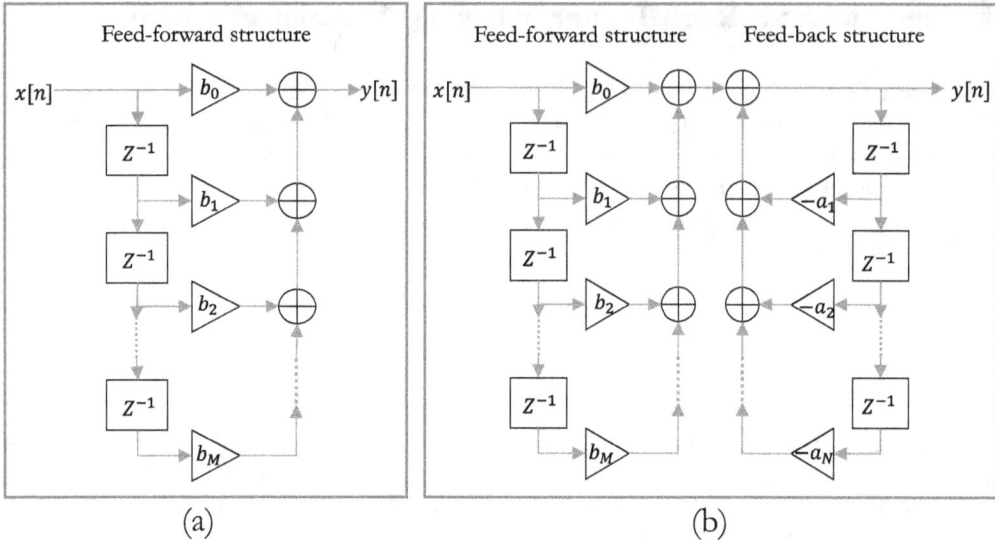

Fig. 1.26: FIR and IIR filters can be realized as direct-form structures. (a) Structure with feed-forward elements only - typical for FIR designs. (b) Structure with feed-back paths - generally results in IIR filters. Other structures like lattice implementations are also available.

1.9.2 General considerations in design

As specified earlier, the choice of filter and the design process depends on design specification, application and the performance issues associates with them. However, the following general considerations are applied in practical design.

Minimizing number of computations

In order to minimize memory requirements for storing the filter co-efficients $\{a_k\},\{b_k\}$ and to minimize the number of computations, ideally we would like $N + M$ to be as small as possible. For the same specification, IIR filters result in much lower order when compared to its FIR counter part. Therefore, IIR filters are efficient when viewed from this standpoint.

Need for real-time processing

The given application may require processing of input samples in real-time or the input samples may exist in a recorded state (example: video/audio playback, image processing applications, audio compression). From this perspective, we have two types of filter systems

- Causal filter

  - Filter output depends on present and past input samples, not on the future samples. The output may also depend on the past output samples, as in IIR filters. Strictly no future samples.
  - Such filters are very much suited for real-time applications.

- Non-causal filter

- There are many practical cases where a non-causal filter is required. Typically, such application warrants some form of post-processing, where the entire data stream is already stored in memory.
- In such cases, a filter can be designed that can take in all type of input samples : present, past and future, for processing. These filters are classified as non-causal filters.
- Non-causal filters have much simpler design methods.

It can be often seen in many signal processing texts, that the causal filters are practically realizable. That does not mean non-causal filters are not practically implementable. In fact both types of filters are implementable and you can see them in many systems today. The question you must ask is : whether your application requires real-time processing or processing of pre-recorded samples. If the application requires *real-time processing*, causal filters must be used. Otherwise, non-causal filters can be used.

Consequences of causal filter

If the application requires real-time processing, causal filters are the only choice for implementation. Following consequences must be considered if causality is desired. Ideal filters with finite bands of zero response (example: brick-wall filters), cannot be implemented using causal filter structure. A direct consequence of causal filter is that the response cannot be ideal. Hence, we must design the filter that provides a close approximation to the desired response . If tolerance specification is given, it has to be met. For example, in the case of designing a low pass filter with given passband frequency (ω_P) and stopband frequencies (ω_S), additional tolerance specifications like allowable passband ripple factor (δ_P), stopband ripple factor (δ_S) need to be considered for the design (Figure 1.27). Therefore, the practical filter design involves choosing ω_P, ω_S, δ_P and δ_S and then designing the filter with $N, M, \{a_k\}$ and $\{b_k\}$ that satisfies all the given requirements/responses. Often, iterative procedures may be required to satisfy all the above (example: Parks and McClellan algorithm used for designing optimal causal FIR filters [17]).

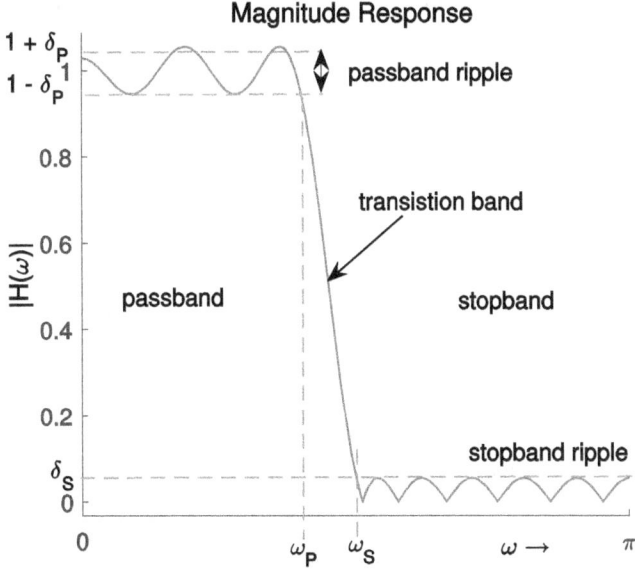

Fig. 1.27: A sample filter design specification

For a causal filter, frequency response's real part $H_R(\omega)$ and the imaginary part $H_I(\omega)$ become *Hilbert transform pair* [18]. Therefore, for a causal filter, the magnitude and phase responses become interdependent.

Stability

A causal LTI digital filter will be *bounded input bounded output*(BIBO) stable, if and only if the impulse response $h[n]$ is absolutely summable.

$$\sum_{n=-\infty}^{n=\infty} |h[n]| < \infty \tag{1.91}$$

Impulse response of FIR filters are always bounded and hence they are inherently stable. On the other hand, an IIR filter may become unstable if not designed properly.

Consider an IIR filter implemented using a floating point processor that has enough accuracy to represent all the coefficients in the following transfer function:

$$H_1(z) = \frac{1}{1 - 1.845\,z^{-1} + 0.850586\,z^{-2}} \tag{1.92}$$

The corresponding impulse response $h_1[n]$ is plotted in Figure 1.28(a). The plot shows that the impulse response decays rapidly to zero as n increases. For this case, the sum in equation 1.91 will be finite. Hence this IIR filter is stable.

Suppose, if we were to implement the same filter in a fixed point processor and we are forced to round-off the co-efficients to 2 digits after the decimal point, the same transfer function looks like this

$$H_2(z) = \frac{1}{1 - 1.85\,z^{-1} + 0.85\,z^{-2}} \tag{1.93}$$

The corresponding impulse response $h_2[n]$ plotted in Figure 1.28(b) implies that the impulse response increases rapidly towards a constant value as n increases. For this case, the sum in equation 1.91 will approach infinity. Hence the implemented IIR filter is unstable.

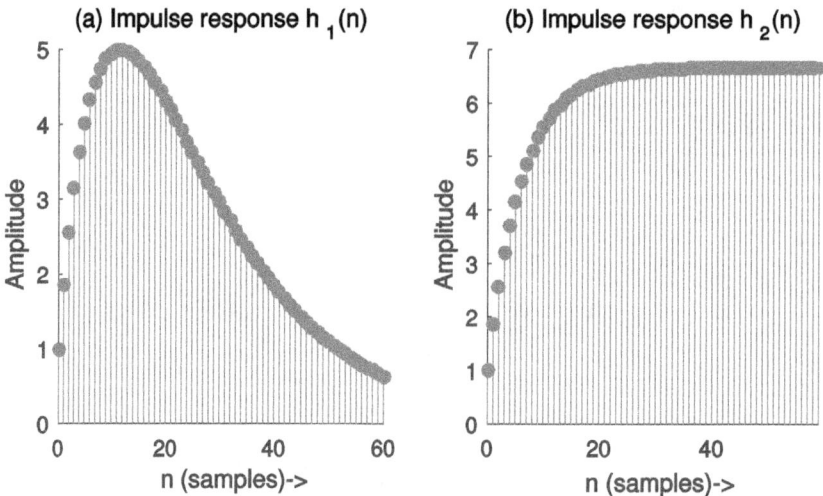

Fig. 1.28: Impact of poorly implemented IIR filter on stability. (a) Stable IIR filter, (b) The same IIR filter becomes unstable due to rounding effects.

Therefore, it is imperative that an IIR filter implementation need to be tested for stability. To analyze the stability of the filter, the infinite sum in equation 1.91 need to be computed and it is often difficult to compute this sum. Analysis of pole-zero plot is an alternative solution for this problem. To have a stable causal filter, the poles of the transfer function should rest completely, strictly, inside the unit circle on the pole-zero plot. The pole-zero plot for the above given transfer functions $H_1(z)$, $H_2(z)$ are plotted in Figure 1.29. It shows that for the transfer function $H_1(z)$, all the poles lie within the unit circle (the region of stability) and hence it is a stable IIR filter. On the other hand, for the transfer function $H_2(z)$, one poles lie exactly on the unit circle (ie., it is just out of the region of stability) and hence it is an unstable IIR filter.

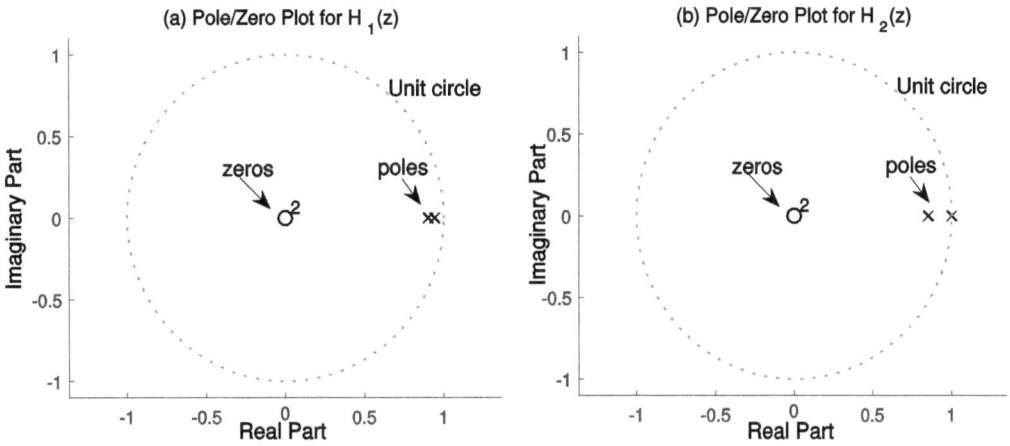

Fig. 1.29: Impact of poorly implemented IIR filter on stability. (a) Stable causal IIR filter since all poles rest inside the unit circle, (b) Due to rounding effects, one of the poles rests on the unit circle, making it unstable.

Linear phase requirement

In many signal processing applications, it is needed that a digital filter should not alter the angular relationship between the real and imaginary components of a signal, especially in the passband. In otherwords, the phase relationship between the signal components should be preserved in the filter's passband. If not, we have phase distortion.

Phase distortion is a concern in many signal processing applications. For example, in phase modulations like GMSK [19],the entire demodulation process hinges on the phase relationship between the inphase and quadrature components of the incoming signal. If we have a phase distorting filter in the demodulation chain, the entire detection process goes for a toss. Hence, we have to pay attention to the phase characteristics of such filters. To have no phase distortion when processing a signal through a filter, every spectral component of the signal inside the passband should be delayed by the same amount time delay measured in samples. In other words, the phase response $\phi(\omega)$ in the passband should be a linear function (straight line) of frequency (except for the phase wraps at the band edges). A filter that satisfies this property is called a *linear phase filter*. FIR filters provide perfect linear phase characteristic in the passband region (Figure 1.30) and hence avoids phase distortion. All IIR filters provide non-linear phase characteristic. If a real-time application warrants for zero phase distortion, FIR filters are the immediate choice for design.

It is intuitive to see the phase response of a generalized linear phase filter should follow the relationship $\phi(\omega) = -m\omega + c$, where m is the slope and c is the intercept when viewing the linear relationship between the frequency and the phase response in the passband (Figure 1.30). The *phase delay* and *group delay* are the important filter characteristics considered for ascertaining the phase distortion and they relate to the intercept c and the slope m of the phase response in the passband. Linear phase filters are characterized by constant

Fig. 1.30: An FIR filter showing linear phase characteristic in the passband

group delay. Any deviation in the group delay from the constant value inside the passband, indicates presence of certain degree of non-linearity in the phase and hence causes phase distortion.

Phase delay is the time delay experienced by each spectral component of the input signal. For a filter with the frequency response $H(\omega)$, the phase delay response τ_p is defined in terms of phase response $\phi(\omega) = \angle H(\omega)$ as

$$\tau_p(\omega) = -\frac{\phi(\omega)}{\omega} \tag{1.94}$$

Group delay is the delay experienced by a group of spectral components within a narrow frequency interval about ω [20]. The group delay response $\tau_g(\omega)$ is defined as the negative derivative of the phase response ω.

$$\tau_g(\omega) = -\frac{d}{d\omega}\phi(\omega) \tag{1.95}$$

For the generalized linear phase filter, the phase delay and group delay are given by

$$\tau_g(\omega) = -\frac{d}{d\omega}\phi(\omega) = m \tag{1.96}$$

$$\tau_p(\omega) = -\frac{\phi(\omega)}{\omega} = m - \frac{c}{\omega} \tag{1.97}$$

Summary of design choices

- IIR filters are efficient, they can provide similar magnitude response for fewer coefficients or lower side-lobes for same number of coefficients
- For linear phase requirement, FIR filters are the immediate choice for the design

- FIR filters are inherently stable. IIR filters are susceptible to finite length words effects of fixed point arithmetic and hence the design has to be rigorously tested for stability.
- IIR filters provide less average delay compared to its equivalent FIR counterpart. If the filter has to be used in a feedback path in a system, the amount of filter delay is very critical as it affects the stability of the overall system.
- Given a specification, an IIR design can be easily deduced based on closed-form expressions. However, satisfying the design requirements using an FIR design, generally requires iterative procedures.

References

1. James W. Cooley and John W. Tukey, *An Algorithm for the Machine Calculation of Complex Fourier Series*, Mathematics of Computation Vol. 19, No. 90, pp. 297-301, April 1965.
2. Good I. J, *The interaction algorithm and practical Fourier analysis*, Journal of the Royal Statistical Society, Series B, Vol. 20, No. 2 , pp. 361-372, 1958.
3. Thomas L. H, *Using a computer to solve problems in physics*, Applications of Digital Computers, Boston, Ginn, 1963.
4. *Matlab documentation on Floating-Point numbers*, https://www.mathworks.com/help/matlab/matlab_prog/floating-point-numbers.html
5. Hanspeter Schmid, *How to use the FFT and Matlab's pwelch function for signal and noise simulations and measurements*, Institute of Microelectronics, University of Applied Sciences Northwestern Switzerland, August 2012.
6. Sanjay Lall, Norm and Vector spaces, Information Systems Laboratory, Stanford. University.http://floatium.stanford.edu/engr207c/lectures/norms_2008_10_07_01.pdf
7. Reddi.S.S, *Eigen Vector properties of Toeplitz matrices and their application to spectral analysis of time series*, Signal Processing, Vol 7, North-Holland, 1984, pp. 46-56.
8. Robert M. Gray, *Toeplitz and circulant matrices – an overview*, Department of Electrical Engineering, Stanford University, Stanford 94305,USA. https://ee.stanford.edu/~gray/toeplitz.pdf
9. Matlab documentation help on Toeplitz command. https://www.mathworks.com/help/matlab/ref/toeplitz.html
10. Richard E. Blahut, *Fast Algorithms for Signal Processing*, Cambridge University Press, 1 edition, August 2010.
11. D. Gabor, *Theory of communications*, Journal of the Inst. Electr. Eng., vol. 93, pt. 111, pp. 42-57, 1946. See definition of complex signal on p. 432.
12. J. A. Ville, *Theorie et application de la notion du signal analytique*, Cables el Transmission, vol. 2, pp. 61-74, 1948.
13. S. M. Kay, *Maximum entropy spectral estimation using the analytical signal*, IEEE transactions on Acoustics, Speech, and Signal Processing, vol. 26, pp. 467-469, October 1978.
14. Poularikas A. D. *Handbook of Formulas and Tables for Signal Processing*, ISBN - 9781420049701, CRC Press , 1999.
15. S. L. Marple, *Computing the discrete-time analytic signal via FFT*, Conference Record of the Thirty-First Asilomar Conference on Signals, Systems and Computers , Pacific Grove, CA, USA, 1997, pp. 1322-1325 vol.2.
16. Moon, Il Joon, and Sung Hwa Hong. *What Is Temporal Fine Structure and Why Is It Important ?*, Korean Journal of Audiology 18.1 (2014): 1–7. PMC. Web. 24 Apr. 2017.
17. J. H. McClellan, T. W. Parks, and L. R. Rabiner, *A Computer Program for Designing Optimum FIR Linear Phase Digital Filters*, IEEE Trans, on Audio and Electroacoustics, Vol. AU-21, No. 6, pp. 506-526, December 1973.
18. Frank R. Kschischang, *The Hilbert Transform*, Department of Electrical and Computer Engineering, University of Toronto, October 22, 2006.
19. Thierry Turletti, *GMSK in a nutshell*, Telemedia Networks and Systems Group Laboratory for Computer Science, Massachussets Institute of Technology April, 1996.
20. Julius O. Smith III, *Introduction to digital filters - with audio applications*, Center for Computer Research in Music and Acoustics (CCRMA), Department of Music, Stanford University.

Chapter 2
Digital Modulators and Demodulators - Passband Simulation Models

Abstract This chapter focuses on the passband simulation models for various modulation techniques including BPSK, differentially encoded BPSK, differential BPSK, QPSK, offset QPSK, pi/4 QPSK, CPM and MSK, GMSK, FSK. Power spectral density (PSD) and performance analysis for these techniques are also provided
.

2.1 Introduction

For a given modulation technique, there are two ways to implement the simulation model: *passband model* and *equivalent baseband model*. The passband model is also called *waveform level simulation model*. The waveform level simulation techniques, described in this chapter, are used to represent the physical interactions of the transmitted signal with the channel. In the waveform level simulations, the transmitted signal, the noise and received signal are all represented by samples of waveforms.

Typically, a waveform level simulation uses many samples per symbol. For the computation of error rate performance of various digital modulation techniques, the value of the symbol at the symbol-sampling time instant is all the more important than the look of the entire waveform. In such a case, the detailed waveform level simulation is not required, instead *equivalent baseband discrete-time model*, described in chapter 3 can be used. Discrete-time equivalent channel model requires only one sample per symbol, hence it consumes less memory and yields results in a very short span of time.

In any communication system, the transmitter operates by modulating the information bearing baseband waveform on to a sinusoidal RF carrier resulting in a *passband* signal. The carrier frequency, chosen for transmission, varies for different applications. For example, FM radio uses $88 - 108 \; MHz$ carrier frequency range, whereas for indoor wireless networks the center frequency of transmission is $1.8 \; GHz$. Hence, the carrier frequency is not the component that contains the information, rather it is the baseband signal that contains the information that is being conveyed.

Actual RF transmission begins by converting the baseband signals to passband signals by the process of up-conversion. Similarly, the passband signals are down-converted to baseband at the receiver, before actual demodulation could begin. Based on this context, two basic types of behavioral models exist for simulation of communication systems - passband models and its baseband equivalent. In the passband model, every cycle of the RF carrier is simulated in detail and the power spectrum will be concentrated near the carrier frequency f_c. Hence, passband models consume more memory, as every point in the RF carrier needs to be stored in computer memory for simulation.

On the other hand, the signals in baseband models are centered near zero frequency. In baseband equivalent models, the RF carrier is suppressed and therefore the number of samples required for simulation is greatly reduced. Furthermore, if the behavior of the system is well understood, the baseband model can be further simplified and the system can be implemented entirely based on the samples at symbol-sampling time instants. Conversion of passband model to baseband equivalent model is discussed in chapter 3 section 3.2.

2.2 Binary Phase Shift Keying (BPSK)

Binary Phase Shift Keying (BPSK) is a two phase modulation scheme, where the 0's and 1's in a binary message are represented by two different phase states in the carrier signal: $\theta = 0°$ for binary 1 and $\theta = 180°$ for binary 0.

In digital modulation techniques, a set of basis functions are chosen for a particular modulation scheme. Generally, the basis functions are orthogonal to each other. Basis functions can be derived using *Gram Schmidt orthogonalization* procedure [1]. Once the basis functions are chosen, any vector in the signal space can be represented as a linear combination of them. In BPSK, only one sinusoid is taken as the basis function. Modulation is achieved by varying the phase of the sinusoid depending on the message bits. Therefore, within a bit duration T_b, the two different phase states of the carrier signal are represented as

$$s_1(t) = A_c \, cos(2\pi f_c t), \qquad 0 \leq t \leq T_b \ \text{ for binary 1}$$
$$s_0(t) = A_c \, cos(2\pi f_c t + \pi), \quad 0 \leq t \leq T_b \ \text{ for binary 0} \qquad (2.1)$$

where, A_c is the amplitude of the sinusoidal signal, f_c is the carrier frequency (Hz), t being the instantaneous time in seconds, T_b is the bit period in seconds. The signal $s_0(t)$ stands for the carrier signal when information bit $a_k = 0$ was transmitted and the signal $s_1(t)$ denotes the carrier signal when information bit $a_k = 1$ was transmitted.

The constellation diagram for BPSK (Figure 2.3) will show two constellation points, lying entirely on the x axis (inphase). It has no projection on the y axis (quadrature). This means that the BPSK modulated signal will have an in-phase component but no quadrature component. This is because it has only one basis function. It can be noted that the carrier phases are 180° apart and it has constant envelope. The carrier's phase contains all the information that is being transmitted.

2.2.1 BPSK transmitter

A BPSK transmitter, shown in Figure 2.1, is implemented by coding the message bits using NRZ coding (1 represented by positive voltage and 0 represented by negative voltage) and multiplying the output by a reference oscillator running at carrier frequency f_c.

Fig. 2.1: BPSK transmitter

The following function (bpsk_mod) implements a *baseband* BPSK transmitter according to Figure 2.1. The output of the function is in baseband and it can optionally be multiplied with the carrier frequency outside the function. In order to get nice continuous curves, the oversampling factor (L) in the simulation should be appropriately chosen. If a carrier signal is used, it is convenient to choose the oversampling factor as the ratio of sampling frequency (f_s) and the carrier frequency (f_c). The chosen sampling frequency must satisfy the Nyquist sampling theorem with respect to carrier frequency. For baseband waveform simulation, the oversampling factor can simply be chosen as the ratio of bit period (T_b) to the chosen sampling period (T_s), where the sampling period is sufficiently smaller than the bit period.

Program 2.1: *bpsk_mod.m*: Baseband BPSK modulator

```
function [s_bb,t] = bpsk_mod(ak,L)
%Function to modulate an incoming binary stream using BPSK(baseband)
%ak - input binary data stream (0's and 1's) to modulate
%L - oversampling factor (Tb/Ts)
%s_bb - BPSK modulated signal(baseband)
%t - generated time base for the modulated signal
N = length(ak); %number of symbols
a = 2*ak-1; %BPSK modulation
ai=repmat(a,1,L).'; %bit stream at Tb baud with rect pulse shape
ai = ai(:).';%serialize
t=0:N*L-1; %time base
s_bb = ai;%BPSK modulated baseband signal
```

2.2.2 BPSK receiver

A correlation type coherent detector, shown in Figure 2.2, is used for receiver implementation. In coherent detection technique, the knowledge of the carrier frequency and phase must be known to the receiver. This can be achieved by using a *Costas loop* or a *Phase Lock Loop (PLL)* at the receiver. For simulation purposes, we simply assume that the carrier phase recovery was done and therefore we directly use the generated reference frequency at the receiver - $cos(2\pi f_c t)$.

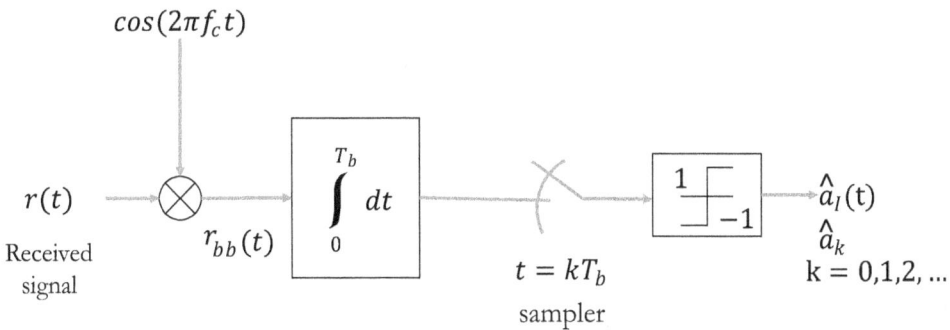

Fig. 2.2: Coherent detection of BPSK (correlation type)

In the coherent receiver, the received signal is multiplied by a reference frequency signal from the carrier recovery blocks like PLL or Costas loop. Here, it is assumed that the PLL/Costas loop is present and the output is completely synchronized. The multiplied output is integrated over one bit period using an integrator. A threshold detector makes a decision on each integrated bit based on a threshold. Since, NRZ signaling format was used in the transmitter, the threshold for the detector would be set to 0. The function bpsk_demod, implements a *baseband* BPSK receiver according to Figure 2.2. To use this function in waveform simulation, first, the received waveform has to be downconverted to baseband, and then the function may be called.

Program 2.2: *bpsk_demod.m*: Baseband BPSK detection (correlation receiver)

```
1  function [ak_cap] = bpsk_demod(r_bb,L)
2  %Function to demodulate an BPSK(baseband) signal
3  %r_bb - received signal at the receiver front end (baseband)
4  %N - number of symbols transmitted
5  %L - oversampling factor (Tsym/Ts)
6  %ak_cap - detected binary stream
7  x=real(r_bb); %I arm
8  x = conv(x,ones(1,L));%integrate for L (Tb) duration
9  x = x(L:L:end);%I arm - sample at every L
10 ak_cap = (x > 0).'; %threshold detector
```

2.2.3 End-to-end simulation

The complete waveform simulation for the end-to-end transmission of information using BPSK modulation is given next. The simulation involves: generating random message bits, modulating them using BPSK modulation, addition of AWGN noise according to the chosen signal-to-noise ratio and demodulating the noisy signal using a coherent receiver. The topic of adding AWGN noise according to the chosen signal-to-noise ratio is discussed in section 4.1 in chapter 4.

The resulting waveform plots are shown in the Figure 2.3. The performance simulation for the BPSK transmitter/receiver combination is also coded in the program shown next (see chapter 4 for more details on theoretical error rates). The resulting performance curves will be same as the ones obtained using the complex baseband equivalent simulation technique in Figure 4.4 of chapter 4.

Program 2.3: *bpsk_wfm_sim.m*: Waveform simulation for BPSK modulation and demodulation

```
1  % Demonstration of BPSK tx/rx chain (waveform simulation)
2  clearvars ; clc;
3  N=100000;%Number of symbols to transmit
4  EbN0dB = -4:2:10; % Eb/N0 range in dB for simulation
5  L=16;%oversampling factor,L=Tb/Ts(Tb=bit period,Ts=sampling period)
6  %if a carrier is used, use L = Fs/Fc, where Fs >> 2xFc
7  Fc=800; %carrier frequency
8  Fs=L*Fc;%sampling frequency
9
10 EbN0lin = 10.^(EbN0dB/10); %converting dB values to linear scale
11 BER = zeros(length(EbN0dB),1); %for SER values for each Eb/N0
12
13 ak = rand(N,1)>0.5; %random symbols from 0's and 1's
14 [s_bb,t]= bpsk_mod(ak,L); %BPSK modulation(waveform) - baseband
15 s = s_bb.*cos(2*pi*Fc*t/Fs); %with carrier
16
```

```
17  %Waveforms at the transmitter
18  subplot(2,2,1);plot(t,s_bb);%baseband wfm zoomed to first 10 bits
19  xlabel('t(s)'); ylabel('s_{bb}(t)-baseband');xlim([0,10*L]);
20  subplot(2,2,2);plot(t,s); %transmitted wfm zoomed to first 10 bits
21  xlabel('t(s)'); ylabel('s(t)-with carrier');xlim([0,10*L]);
22  %signal constellation at transmitter
23  subplot(2,2,3);plot(real(s_bb),imag(s_bb),'o');
24  xlim([-1.5 1.5]); ylim([-1.5 1.5]);
25
26  for i=1:length(EbN0dB),
27      Eb=L*sum(abs(s).^2)/length(s); %signal energy
28      N0= Eb/EbN0lin(i); %Find the noise spectral density
29      n = sqrt(N0/2)*randn(1,length(s));%computed noise
30
31      r = s + n;%received signal with noise
32
33      r_bb = r.*cos(2*pi*Fc*t/Fs);%recovered baseband signal
34      ak_cap = bpsk_demod(r_bb,L);%baseband correlation demodulator
35      BER(i) = sum(ak~=ak_cap)/N;%Symbol Error Rate Computation
36
37      %Received signal waveform zoomed to first 10 bits
38      subplot(2,2,4);plot(t,r);%received signal (with noise)
39      xlabel('t(s)'); ylabel('r(t)');xlim([0,10*L]);
40      pause;%wait for keypress
41  end
42  theoreticalBER = 0.5*erfc(sqrt(EbN0lin));%Theoretical bit error rate
43
44  figure;semilogy(EbN0dB,BER,'k*'); %simulated BER
45  hold on;semilogy(EbN0dB,theoreticalBER,'r-');
46  xlabel('E_b/N_0 (dB)'); ylabel('Probability of Bit Error - P_b');
47  legend('Simulated', 'Theoretical');grid on;
48  title(['Probability of Bit Error for BPSK modulation']);
```

2.3 Coherent detection of Differentially Encoded BPSK (DEBPSK)

In coherent detection, the receiver derives its demodulation frequency and phase references using a carrier synchronization loop. Such synchronization circuits may introduce phase ambiguity $\phi = \hat{\theta} - \theta$ in the detected phase, which could lead to erroneous decisions in the demodulated bits. For example, Costas loop exhibits phase ambiguity of integral multiples of π radians at the lock-in points. As a consequence, the carrier recovery may lock in π radians out-of-phase thereby leading to a situation where all the detected bits are completely inverted when compared to the bits during perfect carrier synchronization. Phase ambiguity can be efficiently combated by applying differential encoding at the BPSK modulator input (Figure 2.4) and by performing differential decoding at the output of the coherent demodulator at the receiver side (Figure 2.5).

In ordinary BPSK transmission, the information is encoded as absolute phases: $\theta = 0°$ for binary 1 and $\theta = 180°$ for binary 0. With differential encoding, the information is encoded as the phase difference between two successive samples. Assuming $a[k]$ is the message bit intended for transmission, the differential encoded output is given as

$$b[k] = b[k-1] \oplus a[k] \quad (modulo-2) \tag{2.2}$$

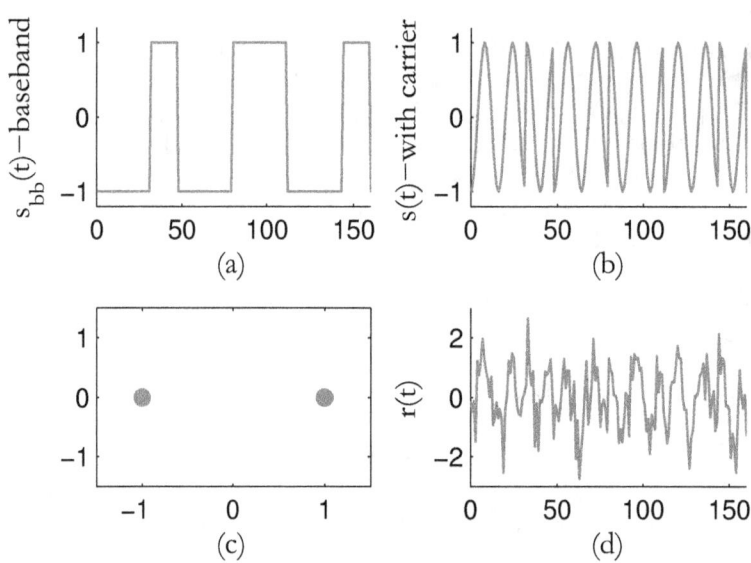

Fig. 2.3: (a) Baseband BPSK signal,(b) transmitted BPSK signal - with carrier, (c) constellation at transmitter and (d) received signal with AWGN noise

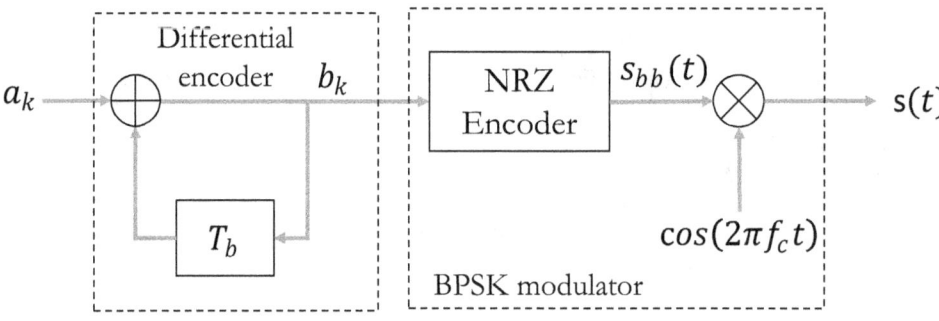

Fig. 2.4: Differential encoded BPSK transmission

The differentially encoded bits are then BPSK modulated and transmitted. On the receiver side, the BPSK sequence is coherently detected and then decoded using a differential decoder. The differential decoding is mathematically represented as

$$a[k] = b[k] \oplus b[k-1] \quad (modulo-2) \tag{2.3}$$

This method can deal with the $180°$ phase ambiguity introduced by synchronization circuits. However, it suffers from performance penalty due to the fact that the differential decoding produces wrong bits when: a) the preceding bit is in error and the present bit is not in error , or b) when the preceding bit is not in error and the present bit is in error. After differential decoding, the average bit error rate of coherently detected BPSK over AWGN channel is given by

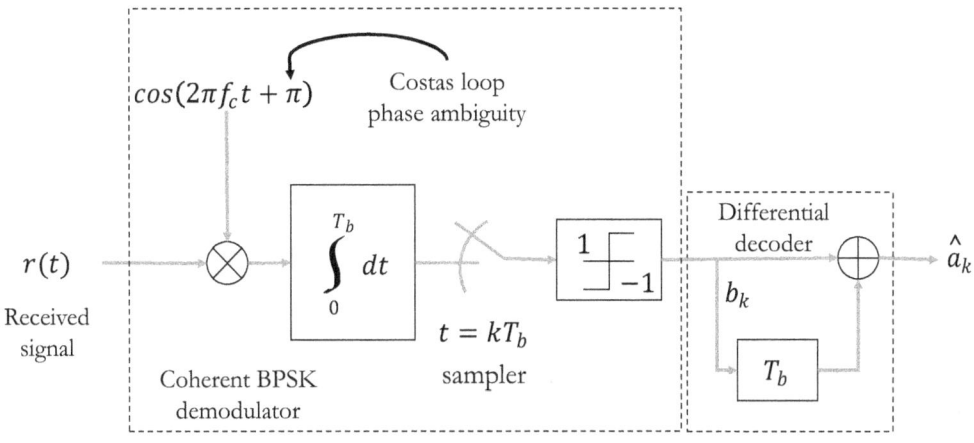

Fig. 2.5: Coherent detection of differentially encoded BPSK signal

$$P_b = erfc\left(\sqrt{\frac{E_b}{N_0}}\right)\left[1 - \frac{1}{2}erfc\left(\sqrt{\frac{E_b}{N_0}}\right)\right] \tag{2.4}$$

Following is the Matlab implementation of the waveform simulation model for the method discussed above. Both the differential encoding and differential decoding blocks, illustrated in Figures2.4 and 2.5), are linear time-invariant filters. The differential encoder is realized using IIR type digital filter and the differential decoder is realized as FIR filter.

Program 2.4: *dbpsk_coherent_detection.m*: Coherent detection of D-BPSK over AWGN channel

```
1   % Coherent detection D-BPSK with phase ambiguity in local oscillator
2   clear all;clc;
3   N=1000000;%Number of symbols to transmit
4   EbN0dB = -4:2:10; % Eb/N0 range in dB for simulation
5   L=16;%oversampling factor,L=Tb/Ts(Tb=bit period,Ts=sampling period)
6   %if a carrier is used, use L = Fs/Fc, where Fs >> 2xFc
7   Fc=800; %carrier frequency
8   Fs=L*Fc;%sampling frequency
9
10  EbN0lin = 10.^(EbN0dB/10); %converting dB values to linear scale
11  EsN0dB=10*log10(1)+EbN0dB; EsN0lin = 10.^(EsN0dB/10);
12  SER = zeros(length(EbN0dB),1);%for SER values for each Eb/N0
13
14  ak = rand(N,1)>0.5; %random symbols from 0's and 1's
15  bk = filter(1,[1 -1],ak,0); %IIR filter for differential encoding
16  bk = mod(bk,2); %XOR operation is equivalent to modulo-2
17  [s_bb,t]= bpsk_mod(bk,L); %BPSK modulation(waveform) - baseband
18  s = s_bb.*cos(2*pi*Fc*t/Fs); %DPSK with carrier
19
20  for i=1:length(EbN0dB),
21      Esym=sum(abs(s).^2)/length(s); %signal energy
22      N0= Esym/EsN0lin(i); %Find the noise spectral density
23      n=sqrt(L*N0/2)*(randn(1,length(s))+1i*randn(1,length(s)));
```

```
24      r = s + n;%received signal with noise
25
26      phaseAmbiguity=pi;%180* phase ambiguity of Costas loop
27      r_bb=r.*cos(2*pi*Fc*t/Fs+phaseAmbiguity);%recoveredsignal
28      b_cap=bpsk_demod(r_bb,L);%baseband correlation type demodulator
29      a_cap=filter([1 1],1,b_cap); %FIR for differential decoding
30      a_cap= mod(a_cap,2); %binary messages, therefore modulo-2
31      SER(i) = sum(ak~=a_cap)/N;%Symbol Error Rate Computation
32
33  end
34  %------Theoretical Bit/Symbol Error Rates------------
35  theorySER_DPSK = erfc(sqrt(EbN0lin)).*(1-0.5*erfc(sqrt(EbN0lin)));
36  theorySER_BPSK = 0.5*erfc(sqrt(EbN0lin));
37  %------------Plots--------------------------
38  figure;semilogy(EbN0dB,SER,'k*'); hold on;
39  semilogy(EbN0dB,theorySER_DPSK,'r-');
40  semilogy(EbN0dB,theorySER_BPSK,'b-');
41  title('Probability of Bit Error for BPSK over AWGN');
42  xlabel('E_b/N_0 (dB)');ylabel('Probability of Bit Error - P_b');
43  legend('Coherent D-BPSK(sim)','Coherent D-BPSK(theory)','Conventional BPSK')
```

Figure 2.6 shows the simulated BER points together with the theoretical BER curves for differentially encoded BPSK and the conventional coherently detected BPSK system over AWGN channel.

Fig. 2.6: Performance of differentially encoded BPSK and conventional BPSK over AWGN channel

2.4 Differential BPSK (D-BPSK)

In section 2.3, differential encoding and coherent detection of binary data was discussed. It was designated as differentially encoded BPSK (DEBPSK). DEBPSK signal can be coherently detected or differentially demodulated (non-coherent detection). In a DEBPSK signal, the information is encoded as the phase difference between two successive samples. The coherent detection method for DEBPSK has not made use of this advantage. If this information can be exploited, the DEBPSK signal can be non-coherently detected.

The differential demodulation of DEBPSK signal is denoted as differential-BPSK or DBPSK, sometimes simply called as DPSK.

2.4.1 Sub-optimum receiver for DBPSK

Coherent detection requires a reference signal with accurate frequency and phase information. However, in non-coherent communications, a coherent reference signal is not available at the receiver. Figure 2.7 illustrates a sub-optimum receiver, which differentially demodulates the DBPSK signal, where the previous symbol serves as a reference to demodulate the current symbol.

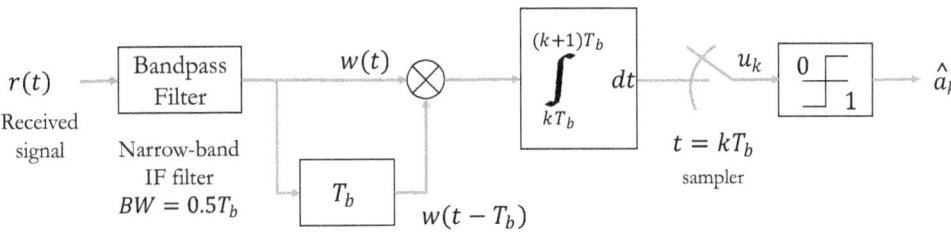

Fig. 2.7: DBPSK sub-optimum Receiver

In the sub-optimum receiver, the received signal is first passed through a narrow-band intermediate (IF) filter with bandwidth set to $BW = 0.5T_b$, where T_b is the bit duration. The IF filter not only reduces the noise in the received signal but also preserves the phase relationships contained in the signal. The filtered output is used to generate a differential signal that drives the integrator. The output of the integrator is given by

$$z = \int_{kT_b}^{(k+1)T_b} r(t)r(t - T_b)dt \tag{2.5}$$

In the absence of noise, the output of the integrator becomes

$$z = \int_{kT_b}^{(k+1)T_b} s_k(t)s_{k-1}(t)dt = \begin{cases} E_b & , if \ s_k(t) = \ s_{k-1}(t) \\ -E_b & , if \ s_k(t) = -s_{k-1}(t) \end{cases} \tag{2.6}$$

where $s_k(t)$ and $s_{k-1}(t)$ are the current and previous transmitted symbols. Thus, the receiver makes decisions based on the difference between two adjacent bits. As a result, a differentially encoded BPSK (DEBPSK) signal can be differentially demodulated using the method above and this receiver structure is termed as *sub-optimum DBPSK receiver*. The bit error probability performance of the sub-optimum receiver is given by [2]

$$P_b = \frac{1}{2}exp\left(-0.76\frac{E_b}{N_0}\right) \tag{2.7}$$

2.4.2 *Optimum noncoherent receiver for DBPSK*

A simpler form of optimum noncoherent receiver for DBPSK is given in Figure 2.8 and its derivation can be found in [3]. In differentially encoded BPSK (DEBPSK), each message bit is represented by two adjacent modulated symbols. If the transmitted bit is 0, the two modulated symbols are same and if the transmitted bit is 1, the two modulated symbols will be different. Since, the phase information is encoded in two adjacent modulated symbols, following signals can be defined for $2T_b$ duration.

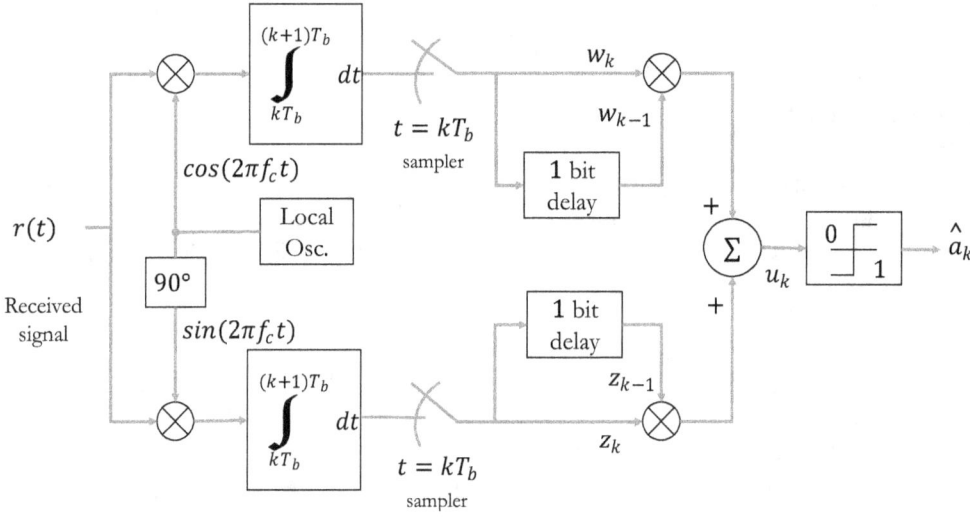

Fig. 2.8: DBPSK optimum noncoherent Receiver

$$\xi_1 = \begin{cases} A\cos(2\pi f_c t), & 0 \leq t \leq T_b \\ A\cos(2\pi f_c t), & T_b \leq t \leq 2T_b \end{cases} \quad , \text{ for binary 0} \tag{2.8}$$

$$\xi_2 = \begin{cases} A\cos(2\pi f_c t), & 0 \leq t \leq T_b \\ -A\cos(2\pi f_c t), & T_b \leq t \leq 2T_b \end{cases} \quad , \text{ for binary 1} \tag{2.9}$$

Then the decision rule is given by

$$a_k \triangleq w_k w_{k-1} + z_k z_{k-1} \underset{1}{\overset{0}{\gtrless}} 0 \tag{2.10}$$

where, the signals w_k, w_{k-1}, z_k and z_{k-1} are defined for the bit $2T_b$ duration as

$$w_0 = \int_0^T r(t)\, A\cos(2\pi f_c t)\, dt$$

$$w_1 = \int_T^{2T} r(t)\, A\cos(2\pi f_c t)\, dt$$

$$z_0 = \int_0^T r(t)\, A\sin(2\pi f_c t)\, dt$$

$$z_1 = \int_T^{2T} r(t)\, A\sin(2\pi f_c t)\, dt \tag{2.11}$$

This method does not require phase synchronization between the transmitter and receiver. However, it does require an accurate frequency reference for the carrier. Any change in the carrier frequency must be tracked and synchronized. Therefore, the sub-optimum receiver is a more practical, less complex implementation but is slightly inferior to the optimum receiver. The error rate performance of the optimum noncoherent DBPSK receiver is given by

$$P_b = \frac{1}{2} exp \left(-\frac{E_b}{N_0} \right) \tag{2.12}$$

The Matlab implementation of DEBPSK modulation and its demodulation using both the sub-optimum receiver (Figure 2.7) and the optimum receiver (Figure 2.8) techniques are given next. Figure 2.9 shows the simulated points for the sub-optimum and the optimum receiver for DBPSK, together with the theoretical curves for conventional BPSK and coherently demodulated DEBPSK schemes.

Program 2.5: *dbpsk_noncoherent_detection.m*: Non-coherent detection of DBPSK

```
1   %Coherent detection D-BPSK with phase ambiguity in local oscillator
2   clearvars;clc;
3   N=100000;%Number of symbols to transmit
4   EbN0dB = -4:2:14; % Eb/N0 range in dB for simulation
5   L=8;%oversampling factor,L=Tb/Ts (Tb=bit period,Ts=sampling period)
6   %if a carrier is used, use L = Fs/Fc, where Fs >> 2xFc
7   Fc=800; %carrier frequency
8   Fs=L*Fc;%sampling frequency
9
10  EbN0lin = 10.^(EbN0dB/10); %converting dB values to linear scale
11  BER_suboptimum = zeros(length(EbN0dB),1);%SER sub-optimum receiver
12  BER_optimum = zeros(length(EbN0dB),1);%SER ptimum receiver
13
14  %----------------Transmitter--------------------
15  a = rand(N,1)>0.5; %random symbols from 0's and 1's
16  b = filter(1,[1 -1],a,1);%IIR filter for differential encoding
17  b = mod(b,2); %XOR operation is equivalent to modulo-2
18  [s_bb,t]= bpsk_mod(b,L); %BPSK modulation(waveform) - baseband
19  s = s_bb.*cos(2*pi*Fc*t/Fs); %DPSK with carrier
20
21  for i=1:length(EbN0dB),
22      %Compute and add AWGN noise
23      Esym=L*sum(abs(s).^2)/length(s); %actual symbol energy
24      N0= Esym/EbN0lin(i); %Find the noise spectral density
25      n=sqrt(N0/2)*(randn(1,length(s))+1i*randn(1,length(s)));%noise
26      %Due to oversampling the signal of noise is sqrt(LN0/2).
27      r = s + n;%received signal with noise
28
29      %----------suboptimum receiver--------------
30      p=real(r).*cos(2*pi*Fc*t/Fs);%demodulate to baseband using BPF
31
32      w0=[p zeros(1,L)];%append L samples on one arm for equal lengths
33      w1=[zeros(1,L) p];%delay the other arm by Tb (L samples)
34      w = w0.*w1;%multiplier
35      z = conv(w,ones(1,L));%integrator from kTb to (K+1)Tb (L samples)
36      u = z(L:L:end-L);%sampler t=kTb
37      a_cap = u.'<0;%decision
```

```
38    BER_suboptimum(i) = sum(a~=a_cap)/N;%BER for suboptimum receiver
39
40    %-----------optimum receiver--------------
41    p=real(r).*cos(2*pi*Fc*t/Fs); %multiply I arm by cos
42    q=imag(r).*sin(2*pi*Fc*t/Fs); %multiply Q arm by sin
43
44    x = conv(p,ones(1,L));%integrate I-arm by Tb duration (L samples)
45    y = conv(q,ones(1,L));%integrate Q-arm by Tb duration (L samples)
46    xk = x(L:L:end);%Sample every Lth sample
47    yk = y(L:L:end);%Sample every Lth sample
48
49    w0 = xk(1:end-1);%non delayed version on I-arm
50    w1 = xk(2:end);%1 bit delay on I-arm
51    z0 = yk(1:end-1);%non delayed version on Q-arm
52    z1 = yk(2:end);%1 bit delay on Q-arm
53
54    u =w0.*w1 + z0.*z1;%decision statistic
55    a_cap=u.'<0;%threshold detection
56    BER_optimum(i) = sum(a(2:end)~=a_cap)/N;%BER for optimum receiver
57  end
58  %------Theoretical Bit/Symbol Error Rates-------------
59  theory_DBPSK_optimum = 0.5.*exp(-EbN0lin);
60  theory_DBPSK_suboptimum = 0.5.*exp(-0.76*EbN0lin);
61  theory_DBPSK_coherent=erfc(sqrt(EbN0lin)).*(1-0.5*erfc(sqrt(EbN0lin)));
62  theory_BPSK_conventional = 0.5*erfc(sqrt(EbN0lin));
63  %-------------Plotting--------------------------
64  figure;semilogy(EbN0dB,BER_suboptimum,'k*','LineWidth',1.0);
65  hold on;semilogy(EbN0dB,BER_optimum,'b*','LineWidth',1.0);
66  semilogy(EbN0dB,theory_DBPSK_suboptimum,'m-','LineWidth',1.0);
67  semilogy(EbN0dB,theory_DBPSK_optimum,'r-','LineWidth',1.0);
68  semilogy(EbN0dB,theory_DBPSK_coherent,'k-','LineWidth',1.0);
69  semilogy(EbN0dB,theory_BPSK_conventional,'b-','LineWidth',1.0);
70  set(gca,'XLim',[-4 12]);set(gca,'YLim',[1E-6 1E0]);
71  set(gca,'XTick',-4:2:12);title('Probability of D-BPSK over AWGN');
72  xlabel('E_b/N_0 (dB)');ylabel('Probability of Bit Error - P_b');
73  legend('DBPSK subopt (sim)','DBPSK opt (sim)','DBPSK subopt (theory)','DBPSK opt (
         theory)','coherent DEBPSK','coherent BPSK');
```

2.5 Quadrature Phase Shift Keying (QPSK)

QPSK is a form of phase modulation technique, in which two information bits (combined as one symbol) are modulated at once, selecting one of the four possible carrier phase shift states. The QPSK signal within a symbol duration T_{sym} is defined as

$$s(t) = A \, cos \left[2\pi f_c t + \theta_n \right] \quad , 0 \le t \le T_{sym}, \, n = 1, 2, 3, 4 \tag{2.13}$$

where the signal phase is given by

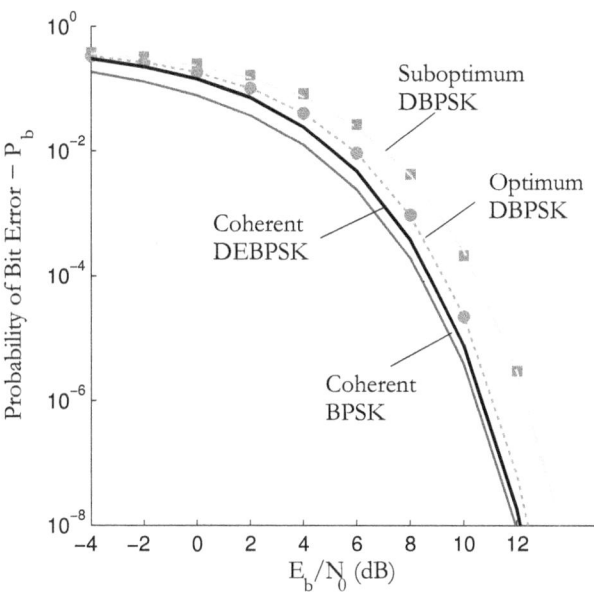

Fig. 2.9: Performance of differential BPSK schemes and coherently detected conventional BPSK

$$\theta_n = (2n-1)\frac{\pi}{4} \tag{2.14}$$

Therefore, the four possible initial signal phases are $\pi/4, 3\pi/4, 5\pi/4$ and $7\pi/4$ radians. Equation 2.13 can be re-written as

$$s(t) = A\cos\theta_n \cos(2\pi f_c t) - A\sin\theta_n \sin(2\pi f_c t) \tag{2.15}$$
$$= s_{ni}\phi_i(t) + s_{nq}\phi_q(t) \tag{2.16}$$

The above expression indicates the use of two orthonormal basis functions: $\langle \phi_i(t), \phi_q(t) \rangle$ together with the inphase and quadrature signaling points: $\langle s_{ni}, s_{nq} \rangle$. Therefore, on a two dimensional co-ordinate system with the axes set to $\phi_i(t)$ and $\phi_q(t)$, the QPSK signal is represented by four constellation points dictated by the vectors $\langle s_{ni}, s_{nq} \rangle$ with $n = 1, 2, 3, 4$.

2.5.1 QPSK transmitter

The QPSK transmitter, shown in Figure 2.10, is implemented as a matlab function qpsk_mod. In this implementation, a splitter separates the odd and even bits from the generated information bits. Each stream of odd bits (quadrature arm) and even bits (in-phase arm) are converted to NRZ format in a parallel manner.

The timing diagram for BPSK and QPSK modulation is shown in Figure 2.11. For BPSK modulation the symbol duration for each bit is same as bit duration, but for QPSK the symbol duration is twice the bit duration: $T_{sym} = 2T_b$. Therefore, if the QPSK symbols were transmitted at same rate as BPSK, it is clear that QPSK sends twice as much data as BPSK does. After oversampling and pulse shaping, it is intuitively clear that the signal on the I-arm and Q-arm are BPSK signals with symbol duration $2T_b$. The signal on the in-phase arm is then multiplied by $cos(2\pi f_c t)$ and the signal on the quadrature arm is multiplied by $-sin(2\pi f_c t)$. QPSK modulated signal is obtained by adding the signal from both in-phase and quadrature arms.

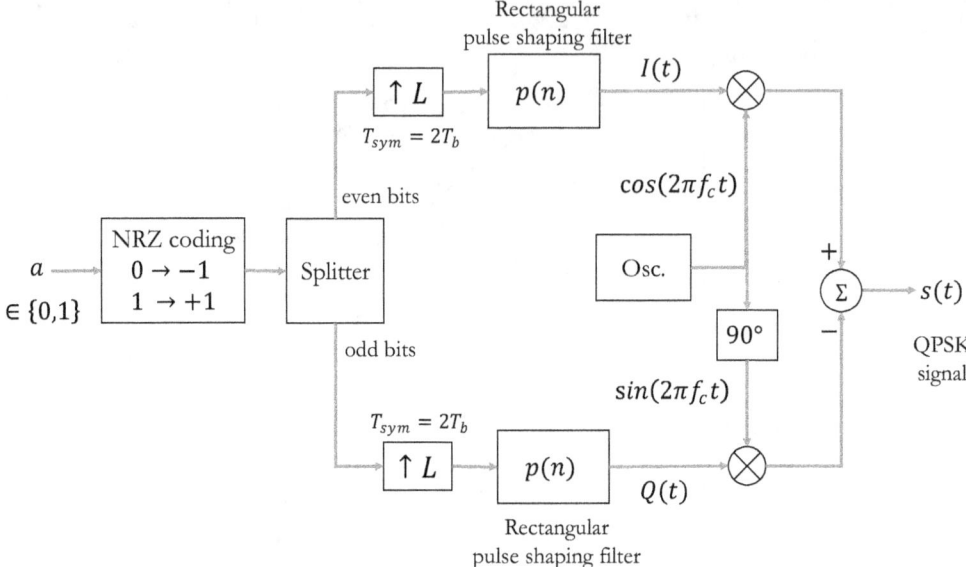

Fig. 2.10: Waveform simulation model for QPSK modulation

Note: The oversampling rate for the simulation is chosen as $L = 2f_s/f_c$, where f_c is the given carrier frequency and f_s is the sampling frequency satisfying Nyquist sampling theorem with respect to the carrier frequency ($f_s \geq fc$). This configuration gives integral number of carrier cycles for one symbol duration.

Program 2.6: *qpsk_mod.m*: QPSK modulator

```
function [s,t,I,Q] = qpsk_mod(a,fc,OF)
%Modulate an incoming binary stream using conventional QPSK
%a - input binary data stream (0's and 1's) to modulate
%fc - carrier frequency in Hertz
%OF - oversampling factor (multiples of fc) - at least 4 is better
%s - QPSK modulated signal with carrier
%t - time base for the carrier modulated signal
%I - baseband I channel waveform (no carrier)
%Q - baseband Q channel waveform (no carrier)
L = 2*OF;%samples in each symbol (QPSK has 2 bits in each symbol)

ak = 2*a-1; %NRZ encoding 0-> -1, 1->+1
I = ak(1:2:end);Q = ak(2:2:end);%even and odd bit streams
I=repmat(I,1,L).'; Q=repmat(Q,1,L).';%even/odd streams at 1/2Tb baud
I = I(:).'; Q = Q(:).'; %serialize

fs = OF*fc; %sampling frequency
t=0:1/fs:(length(I)-1)/fs; %time base
iChannel = I.*cos(2*pi*fc*t);qChannel = -Q.*sin(2*pi*fc*t);
s = iChannel + qChannel; %QPSK modulated baseband signal

doPlot=1; %set to 0 if you do not intend to see waveform plots
if doPlot==1,%Waveforms at the transmitter
```

```
24   figure;subplot(3,2,1);plot(t,I);%first few symbols of baseband I
25   xlabel('t'); ylabel('I(t)-baseband');xlim([0,10*L/fs]);
26   subplot(3,2,2);plot(t,Q);%first few symbols of baseband Q signal
27   xlabel('t'); ylabel('Q(t)-baseband');xlim([0,10*L/fs]);
28   subplot(3,2,3);plot(t,iChannel,'r');%I(t) with carrier
29   xlabel('t'); ylabel('I(t)-with carrier');xlim([0,10*L/fs]);
30   subplot(3,2,4);plot(t,qChannel,'r');%Q(t) with carrier
31   xlabel('t'); ylabel('Q(t)-with carrier');xlim([0,10*L/fs]);
32   subplot(3,1,3);plot(t,s); %QPSK waveform zoomed to first few symbols
33   xlabel('t'); ylabel('s(t)');xlim([0,10*L/fs]);
34   end
```

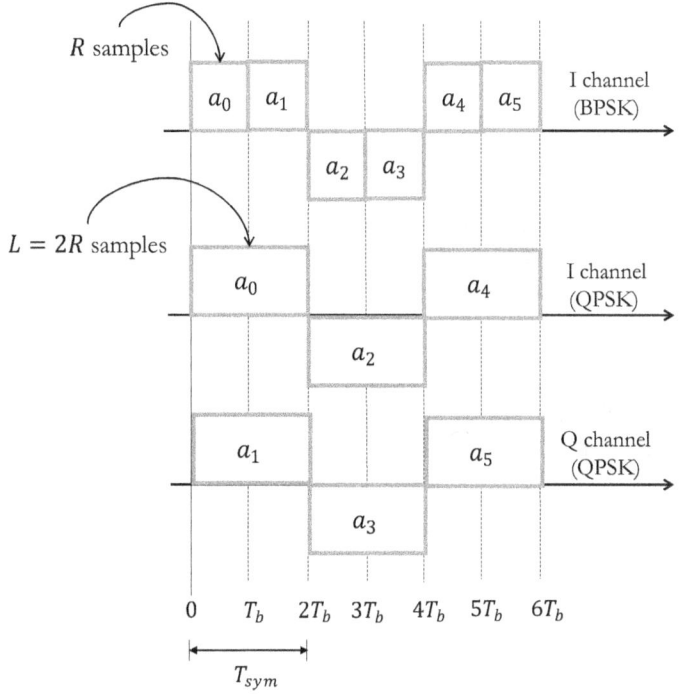

Fig. 2.11: Timing diagram for BPSK and QPSK modulations

2.5.2 QPSK receiver

Due to its special relationship with BPSK, the QPSK receiver takes the simplest form as shown in Figure 2.12. In this implementation, the I-channel and Q-channel signals are individually demodulated in the same way as that of BPSK demodulation. After demodulation, the I-channel bits and Q-channel sequences are combined into a single sequence. The function qpsk_demod implements a QPSK demodulator as per Figure 2.12

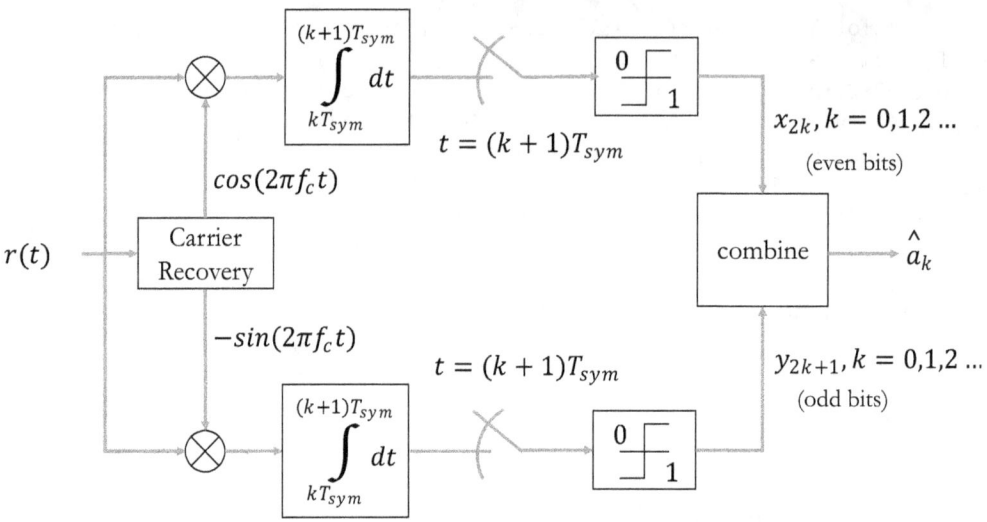

Fig. 2.12: Waveform simulation model for QPSK demodulation

Program 2.7: *qpsk_demod.m*: QPSK demodulator

```
function [a_cap] = qpsk_demod(r,fc,OF)
%Function to demodulate a conventional QPSK signal
%r - received signal at the receiver front end
%fc - carrier frequency in Hertz
%OF - oversampling factor (multiples of fc) - at least 4 is better
%L - upsampling factor on the inphase and quadrature arms
%a_cap - detected binary stream
fs = OF*fc; %sampling frequency
L = 2*OF; %samples in 2Tb duration
t=0:1/fs:(length(r)-1)/fs; %time base

x=r.*cos(2*pi*fc*t); %I arm
y=-r.*sin(2*pi*fc*t); %Q arm
x = conv(x,ones(1,L));%integrate for L (Tsym=2*Tb) duration
y = conv(y,ones(1,L));%integrate for L (Tsym=2*Tb) duration
x = x(L:L:end);%I arm - sample at every symbol instant Tsym
y = y(L:L:end);%Q arm - sample at every symbol instant Tsym

a_cap = zeros(1,2*length(x));
a_cap(1:2:end) = x.' > 0; %even bits
a_cap(2:2:end) = y.' > 0; %odd bits

doPlot=1; %To plot constellation at the receiver
if doPlot==1, figure; plot(x(1:200),y(1:200),'o'); end
```

2.5.3 Performance simulation over AWGN

The complete waveform simulation for the aforementioned QPSK modulation and demodulation is given next. The simulation involves, generating random message bits, modulating them using QPSK modulation, addition of AWGN channel noise corresponding to the given signal-to-noise ratio and demodulating the noisy signal using a coherent QPSK receiver. The waveforms at the various stages of the modulator are shown in the Figure 2.13. The ideal-constellation at the transmitter and the constellation at the receiver that is affected by channel noise (shown for $E_b/N_0 = 10\ dB$) are shown in Figure 2.29 respectively.

Program 2.8: *qpsk_wfm_sim.m*: Waveform simulation for QPSK modulation and demodulation

```
1   % Demonstration of Eb/N0 Vs BER for QPSK (waveform simulation)
2   clear all;clc;
3   N=100000;%Number of bits to transmit
4   EbN0dB = -4:2:10; % Eb/N0 range in dB for simulation
5   fc=100;%carrier frequency in Hertz
6   OF =8; %oversampling factor, sampling frequency will be fs=OF*fc
7
8   EbN0lin = 10.^(EbN0dB/10); %converting dB values to linear scale
9   BER = zeros(length(EbN0dB),1); %For BER values for each Eb/N0
10
11  a = rand(N,1)>0.5; %random bits - input to QPSK
12  [s,t] = qpsk_mod(a,fc,OF);%QPSK modulation
13
14  for i=1:length(EbN0dB),
15      Eb=OF*sum(abs(s).^2)/(length(s)); %compute energy per bit
16      N0= Eb/EbN0lin(i); %required noise spectral density from Eb/N0
17      n = sqrt(N0/2)*(randn(1,length(s)));%computed noise
18      r = s + n;%add noise
19      a_cap = qpsk_demod(r,fc,OF); %QPSK demodulation
20      BER(i) = sum(a~=a_cap.')/N;%Bit Error Rate Computation
21  end
22  %------Theoretical Bit Error Rate------------
23  theoreticalBER = 0.5*erfc(sqrt(EbN0lin));%Theoretical bit error rate
24  %------------Plot performance curve----------------------
25  figure;semilogy(EbN0dB,BER,'k*','LineWidth',1.5); %simulated BER
26  hold on; semilogy(EbN0dB,theoreticalBER,'r-','LineWidth',1.5);
27  title('Probability of Bit Error for QPSK modulation');
28  xlabel('E_b/N_0 (dB)');ylabel('Probability of Bit Error - P_b');
29  legend('Simulated', 'Theoretical');grid on;
```

The performance simulation for the QPSK transmitter-receiver combination was also coded in the code given above and the resulting bit-error rate performance curve will be same as that of conventional BPSK (Figure 2.6). A QPSK signal essentially combines two orthogonally modulated BPSK signals. Therefore, the resulting performance curves for QPSK - E_b/N_0 Vs. bits-in-error - will be same as that of conventional BPSK.

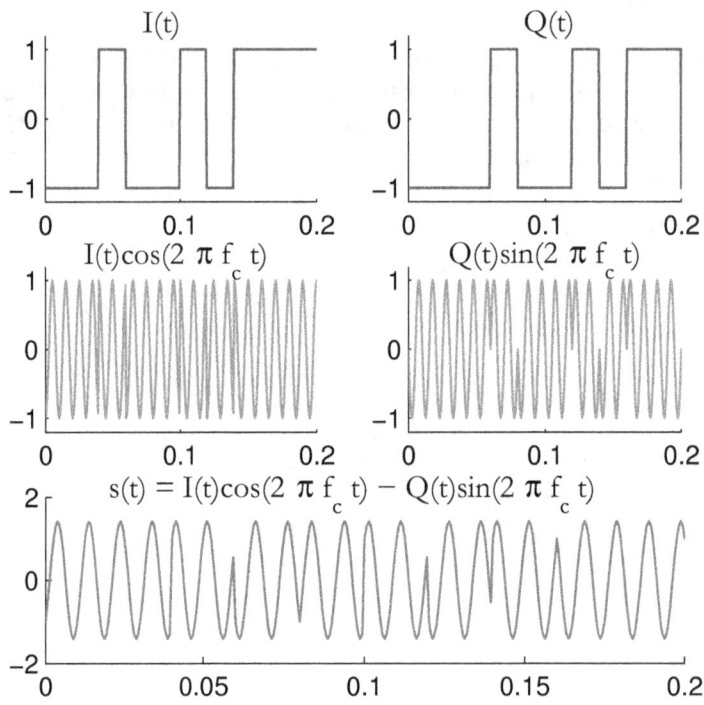

Fig. 2.13: Simulated QPSK waveforms at the transmitter side

2.6 Offset QPSK (O-QPSK)

Offset QPSK is essentially same as QPSK, except that the orthogonal carrier signals on the I-channel and the Q-channel are staggered (one of them is delayed in time). OQPSK modulator and demodulator are shown in Figures 2.14 and 2.15 respectively. From the modulator, shown in Figure 2.14, it is clear that this scheme is same as QPSK except for an extra delay of half symbol period $(T_{sym}/2 = T_b)$ in the Q-channel arm. OQPSK is also referred as staggered QPSK. Based on the modulator, an OQPSK signal can be written as

$$s(t) = I(t)\cos(2\pi f_c t) - Q\left(t - \frac{T_{sym}}{2}\right)\sin(2\pi f_c t) \tag{2.17}$$

The staggering concept in OQPSK is illustrated in Figure 2.16. Since the OQPSK differs from QPSK only in the time alignment of the orthogonal bit streams, both QPSK and OQPSK have the same power spectral density (refer the simulated PSD in Figure 2.30)and same error rate performance. However, these two modulations behave differently when passed through a bandlimited filter as encountered in certain applications like satellite communications. This difference in the behavior is due to the nature of phase changes in the carrier for these two modulations.

In QPSK, due to coincident alignment of I and Q channels, the carrier phase changes at every symbol boundary, i.e, for every symbol duration $(T_{sym} = 2T_b)$. If any one of the I-channel or Q-channel component changes sign, a phase shift of $\pm 90°$ occurs. If both the components change sign, it results in a phase shift of 180°. Phase shifts of 180° in QPSK degrades its constant envelop property and it can cause deleterious effects like out-of-band radiation when passed through bandlimited filters. On the otherhand, OQPSK signal does not suffer from these pitfalls.

Fig. 2.14: OQPSK modulator

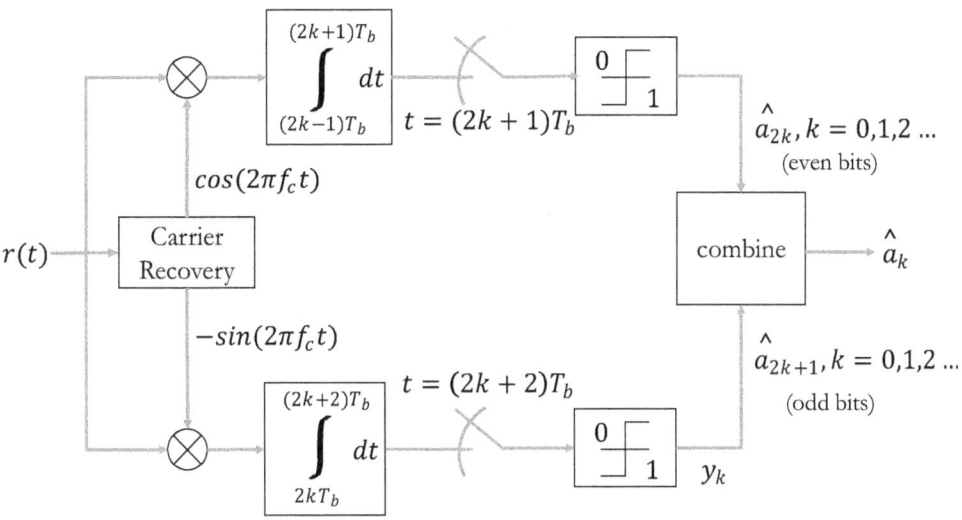

Fig. 2.15: OQPSK demodulator

In OQPSK, the orthogonal components cannot change states at the same time, this is because the components change state only at the middle of the symbol periods (due to the half symbol offset in the Q-channel). This eliminates $180°$ phase shifts all together and the phase changes are limited to $0°$ or $\pm90°$ every T_b seconds.

Elimination of $180°$ phase shifts in OQPSK offers many advantages over QPSK. Unlike QPSK, the spectrum of OQPSK remains unchanged when bandlimited [4]. Additionally, OQPSK performs better than QPSK when subjected to phase jitters [5]. Further improvements to OQPSK can be obtained if the phase transitions are avoided altogether - as evident from continuous modulation schemes like Minimum Shift Keying (MSK) technique.

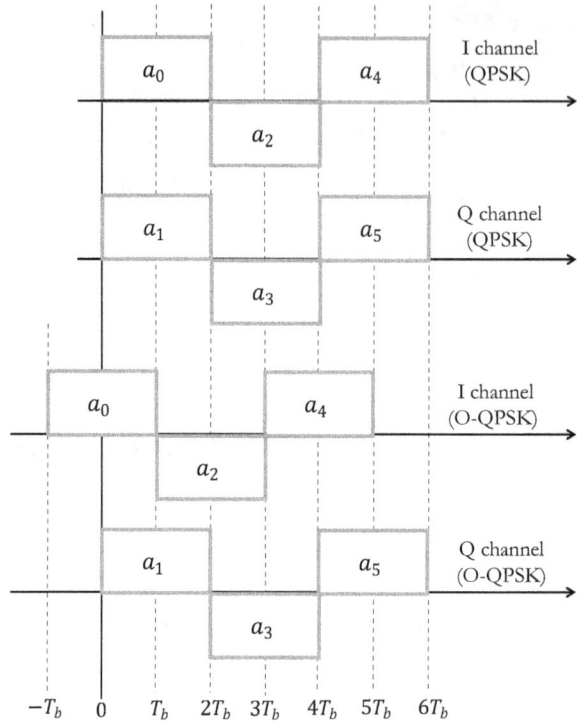

Fig. 2.16: Timing diagrams comparing QPSK and OQPSK modulations

The following functions implement the OQPSK modulator and demodulator given in Figures 2.14 and 2.15 respectively.

Program 2.9: *oqpsk_mod.m*: OQPSK modulator

```
1  function [s,t,I,Q] = oqpsk_mod(a,fc,OF)
2  %Modulate an incoming binary stream using OQPSK (bandpass)
3  %a - input binary data stream (0's and 1's) to modulate
4  %fc - carrier frequency in Hertz
5  %OF - oversampling factor
6  %s - O-QPSK modulated signal with carrier
7  %t - time base for the carrier modulated signal
8  %I - baseband I channel waveform(no carrier)
9  %Q - baseband Q channel waveform(no carrier)
10 L = 2*OF; %samples in each symbol (QPSK has 2 bits per symbol)
11
12 ak = 2*a-1; %NRZ encoding 0-> -1, 1->+1
13 I = ak(1:2:end);Q = ak(2:2:end);%even and odd bit streams
14 I=repmat(I,1,L).'; Q=repmat(Q,1,L).';%even odd streams at 1/2Tb baud
15 I = [I(:) ; zeros(L/2,1)].'; %padding at end to equal length of Q
16 Q = [zeros(L/2,1); Q(:)].'; %delay by Tb (delay by L/2)
17
18 fs = OF*fc; %sampling frequency
19 t=(0:1:length(I)-1)/fs; %time base
```

```
20  iChannel = I.*cos(2*pi*fc*t);qChannel = -Q.*sin(2*pi*fc*t);
21  s = iChannel + qChannel; %O-QPSK modulated baseband signal
22
23  doPlot=1; %set to 0 if you do not intend to see waveform plots
24  if doPlot==1,%Waveforms at the transmitter
25  figure;subplot(3,2,1);plot(t,I);%baseband waveform on I arm
26  xlabel('t'); ylabel('I(t)-baseband');xlim([0,10*L/fs]);
27  subplot(3,2,2);plot(t,iChannel,'r');%I(t) with carrier
28  xlabel('t'); ylabel('I(t)-with carrier');xlim([0,10*L/fs]);
29  subplot(3,2,3);plot(t,Q);%baseband waveform on Q arm
30  xlabel('t'); ylabel('Q(t)-baseband');xlim([0,10*L/fs]);
31  subplot(3,2,4);plot(t,qChannel,'r');%Q(t) with carrier
32  xlabel('t'); ylabel('Q(t)-with carrier');xlim([0,10*L/fs]);
33  subplot(3,1,3);plot(t,s); %QPSK waveform zoomed to first few symbols
34  xlabel('t'); ylabel('s(t)');xlim([0,10*L/fs]);
35  figure;plot(I,Q);axis([-1.5 1.5 -1.5 1.5])%constellation plots at Tx
36  end
```

Program 2.10: *oqpsk_demod.m*: OQPSK demodulator

```
1   function [a_cap] = oqpsk_demod(r,N,fc,OF)
2   %Function to demodulate a bandpass O-QPSK signal
3   %r - received signal at the receiver front end
4   %N - number of symbols transmitted
5   %fc - carrier frequency in Hertz
6   %OF - oversampling factor (multiples of fc) - at least 4 is better
7   %L - upsampling factor on the inphase and quadrature arms
8   %a_cap - detected binary stream
9   fs = OF*fc; %sampling frequency
10  L = 2*OF; %samples in 2Tb duration
11  t=(0:1:(N+1)*OF-1)/fs; %time base
12
13  x=r.*cos(2*pi*fc*t); %I arm
14  y=-r.*sin(2*pi*fc*t); %Q arm
15  x = conv(x,ones(1,L));%integrate for L (Tsym=2*Tb) duration
16  y = conv(y,ones(1,L));%integrate for L (Tsym=2*Tb) duration
17  x = x(L:L:end-L);%I arm - sample at the end of every symbol
18  %Q arm - sample at end of every symbol starting L+L/2th sample
19  y = y(L+L/2:L:end-L/2);
20
21  a_cap = zeros(N,1);
22  a_cap(1:2:end) = x.' > 0; %even bits
23  a_cap(2:2:end) = y.' > 0; %odd bits
24
25  doPlot=0; %To plot constellation at the receiver
26  if doPlot==1, figure; plot(x(1:200),y(1:200),'o');end
```

The wrapper code for simulating a complete communication chain with OQPSK modulator (with carrier), AWGN channel and OQPSK demodulator is given next. The code also includes performance simulation of the OQPSK communication system over an AWGN channel and the performance is essentially same as that of QPSK. The simulated timing waveforms at the transmitter is shown in Figure 2.17.

Program 2.11: *oqpsk_wfm_sim.m*: Waveform simulation for OQPSK modulation and demodulation

```matlab
% Demonstration of Eb/N0 Vs SER for baseband MSK/OQPSK schemes
clear all;clc;
N=100000;%Number of symbols to transmit
EbN0dB = -4:2:10; % Eb/N0 range in dB for simulation
fc=100;%carrier frequency in Hertz
OF =16; %oversampling factor, sampling frequency will be fs=OF*fc

EbN0lin = 10.^(EbN0dB/10); %converting dB values to linear scale
BER = zeros(length(EbN0dB),1); %SER values for each Eb/N0

a = rand(N,1)>0.5; %random symbols from 0's and 1's - input to QPSK
[s,t] = oqpsk_mod(a,fc,OF);%OQPSK modulation

for i=1:length(EbN0dB),
    Eb=OF*sum(abs(s).^2)/(length(s)); %compute energy per bit
    N0= Eb/EbN0lin(i); %required noise spectral density from Eb/N0
    n = sqrt(N0/2)*(randn(1,length(s)));%computed noise
    r = s + n;
    a_cap = oqpsk_demod(r,N,fc,OF);%OQPSK demodulation
    BER(i) = sum(a~=a_cap)/N;%Symbol Error Rate Computation
end
%------Theoretical Bit Error Rate-------------
theoreticalBER = 0.5*erfc(sqrt(EbN0lin));%Theoretical BER
%------------Plot performance curve----------------------
figure;semilogy(EbN0dB,BER,'k*','LineWidth',1.5); %simulated BER
hold on; semilogy(EbN0dB,theoreticalBER,'r-','LineWidth',1.5);
title('Probability of Bit Error for QPSK modulation');
xlabel('E_b/N_0 (dB)');ylabel('Probability of Bit Error - P_b');
legend('Simulated', 'Theoretical');grid on;
```

2.7 $\pi/4$-DQPSK

QPSK modulation has several variants, two such flavors among them are: $\pi/4$-QPSK and $\pi/4$-DQPSK.

In $\pi/4$-QPSK, the signaling points of the modulated signals are chosen from two QPSK constellations that are just shifted $\pi/4$ radians (45°) with respect to each other. Switching between the two constellations every successive bit ensures that the phase changes are confined to odd multiples of 45°. Therefore, phase transitions of $\pm 90°$ and 180° are eliminated.

$\pi/4$-QPSK preserves the constant envelope property better than QPSK and OQPSK. Unlike QPSK and OQPSK schemes, $\pi/4$-QPSK can be differentially encoded therefore enabling the use of both coherent and non-coherent demodulation techniques. Choice of non-coherent demodulation results in simpler receiver design. Differentially encoded $\pi/4$-QPSK is called $\pi/4$-*DQPSK* which is the subject of this section.

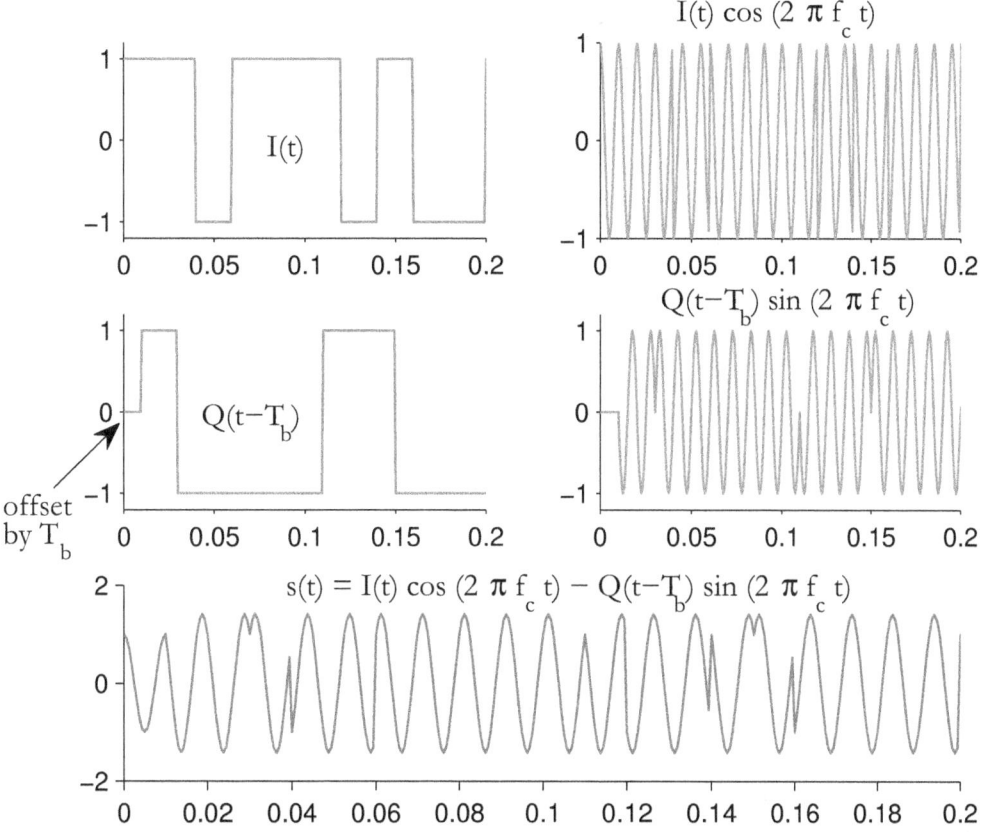

Fig. 2.17: Simulated OQPSK waveforms at the transmitter side

Modulator

The modulator for $\pi/4$-DQPSK is shown in Figure 2.18. The differential encoding for $\pi/4$-DQPSK is defined by the following encoding rules.

Let $\langle I(t), Q(t) \rangle$ and $\langle u(t), v(t) \rangle$ be the unencoded and differentially encoded versions of I-channel and Q-channel waveform sequences. From reference [3], the differentially encoded $\pi/4$-DQPSK signals are obtained as

$$u_k = u_{k-1}\,cos\Delta\theta_k - v_{k-1}\,sin\Delta\theta_k$$
$$v_k = u_{k-1}\,sin\Delta\theta_k + v_{k-1}\,cos\Delta\theta_k \tag{2.18}$$

where, $\Delta\theta_k$ defines the phase difference determined by the input data as given in Table 2.1. The initial values for the very first encoded samples are defined as $u_{-1} = 1$ and $v_{-1} = 0$. After applying the encoding rule given above, the differentially encoded I-channel and Q-channel bits are modulated in the same fashion as the conventional QPSK scheme.

The following Matlab function - piBy4_dqpsk_diff_encoding, implements the encoding rules for the differential encoder. The next shown function - piBy4_dqpsk_mod, implements the $\pi/4$-DQPSK modulator, as per Figure 2.18.

Table 2.1: Phase mapper for $\pi/4$-DQPSK

$I_k\, Q_k$	$\Delta\theta_k$	$cos\Delta\theta_k$	$sin\Delta\theta_k$
$-1\ -1$	$-3\pi/4$	$-1/\sqrt{2}$	$-1/\sqrt{2}$
$-1\ \ \ 1$	$3\pi/4$	$-1/\sqrt{2}$	$1/\sqrt{2}$
$1\ -1$	$-\pi/4$	$1/\sqrt{2}$	$-1/\sqrt{2}$
$1\ \ \ 1$	$\pi/4$	$1/\sqrt{2}$	$1/\sqrt{2}$

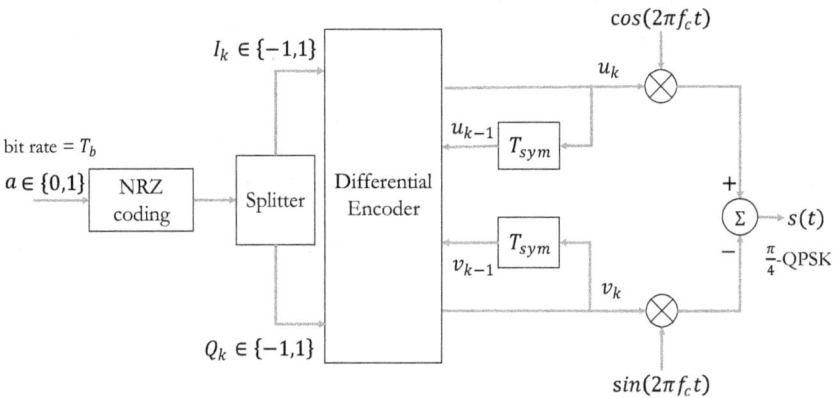

Fig. 2.18: $\pi/4$-DQPSK modulator

Program 2.12: *piBy4_dqpsk_diff_encoding.m*: Differential encoding for $\pi/4 - DQPSK$ - equation 2.18

```matlab
function [ u,v ] = piBy4_dqpsk_Diff_encoding(a)
%Phase Mapper for pi/4-DQPSK modulation
% a - input stream of binary bits
% u - differentially coded I-channel bits
% v - differentially coded Q-channel bits

if mod(length(a),2), error('Length of binary stream must be even');end
I = a(1:2:end);%odd bit stream
Q = a(2:2:end);%even bit stream
%club 2-bits to form a symbol and use it as index for dTheta table
m = 2*I+Q;
dTheta = [-3*pi/4, 3*pi/4, -pi/4, pi/4]; %LUT for pi/4-DQPSK

u = zeros(length(m)+1,1);v = zeros(length(m)+1,1);
u(1)=1; v(1)=0;%initial conditions for uk and vk

for k=1:length(m),
    u(k+1) = u(k) * cos(dTheta(m(k)+1)) - v(k) * sin(dTheta(m(k)+1));
    v(k+1) = u(k) * sin(dTheta(m(k)+1)) + v(k) * cos(dTheta(m(k)+1));
end
figure; plot(v,u);%constellation plot
end
```

Program 2.13: *piBy4_dqpsk_mod.m* : $\pi/4 - DQPSK$ modulator

```matlab
function [s,t,U,V] = piBy4_dqpsk_mod( a,fc,OF )
%Perform pi/4 DQPSK modulation
%a - input binary data stream (0's and 1's) to modulate
%fc - carrier frequency in Hertz
%OF - oversampling factor
%s - pi/4-DQPSK modulated signal with carrier
%t - time base for the carrier modulated signal
%U - differentially coded I-channel waveform (no carrier)
%V - differentially coded Q-channel waveform (no carrier)

[u,v]=piBy4_dqpsk_Diff_encoding(a);%Differential Encoding for pi/4 QPSK
%Waveform formation (similar to conventional QPSK)
L = 2*OF; %number of samples in each symbol (QPSK has 2 bits/symbol)
U=repmat(u,1,L).'; %odd bit stream at 1/2Tb baud
V=repmat(v,1,L).';%even bit stream at 1/2Tb baud
U = U(:).'; %serialize signal on I arm
V = V(:).'; %serialize signal on Q arm

fs = OF*fc; %sampling frequency
t=0:1/fs:(length(U)-1)/fs; %time base
iChannel = U.*cos(2*pi*fc*t);
qChannel = -V.*sin(2*pi*fc*t);
s = iChannel + qChannel;

doPlot=1; %set to 0 if you do not intend to see waveform plots
if doPlot==1,%Waveforms at the transmitter
figure;
subplot(3,2,1);plot(t,U);%zoomed baseband wfm on I arm
xlabel('t'); ylabel('U(t)-baseband');xlim([0,10*L/fs]);
subplot(3,2,2);plot(t,V);%zoomed baseband wfm on Q arm
xlabel('t'); ylabel('V(t)-baseband');xlim([0,10*L/fs]);
subplot(3,2,3);plot(t,iChannel,'r');%U(t) with carrier
xlabel('t'); ylabel('U(t)-with carrier');xlim([0,10*L/fs]);
subplot(3,2,4);plot(t,qChannel,'r');%V(t) with carrier
xlabel('t'); ylabel('V(t)-with carrier');xlim([0,10*L/fs]);
subplot(3,1,3);plot(t,s); %QPSK waveform zoomed to first few symbols
xlabel('t'); ylabel('s(t)');xlim([0,10*L/fs]);
end
end
```

Demodulator

From the equations for $\pi/4$-DQPSK modulation, we can see that the information is encoded as phase difference $\Delta\theta_k$. Therefore, in addition to coherent detection, non-coherent detection techniques can also be used for demodulating the $\pi/4$-DQPSK signal. Of the several demodulation techniques available for demodulating the $\pi/4$-DQPSK signal, the baseband differential detection is used here for simulation. Refer [3] for details on the rest of the demodulator techniques.

The demodulator for baseband differential detection technique is presented in Figure 2.19. Let w_k and z_k be the baseband versions of differentially encoded I-channel and Q-channel bits at the receiver (they are essentially the output of the symbol rate samplers shown in the Figure 2.19). The decoding rules for obtaining the corresponding differentially decoded versions are given by

$$x_k = w_k w_{k-1} + z_k z_{k-1}$$
$$y_k = z_k w_{k-1} - w_k z_{k-1}$$

$$(2.19)$$

The I-channel and Q-channel bits are then decided as

$$\hat{I}_k \triangleq x_k \overset{1}{\underset{0}{\gtrless}} 0 \qquad\qquad (2.20)$$

$$\hat{Q}_k \triangleq y_k \overset{1}{\underset{0}{\gtrless}} 0 \qquad\qquad (2.21)$$

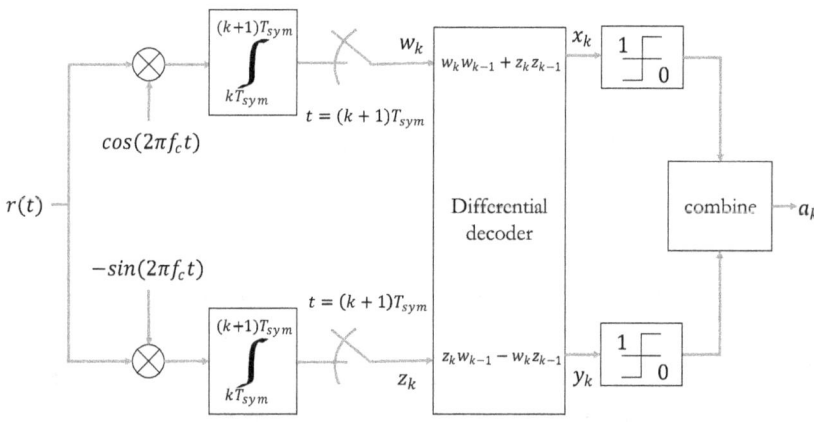

Fig. 2.19: Baseband differential demodulator for $\pi/4$-DQPSK

The Matlab function (piBy4_dqpsk_diff_decoding) implementing the above mentioned equations for the differential decoder is given next. The function implementing the $\pi/4$-DQPSK demodulator as per Figure 2.19, is given next.

Program 2.14: *piBy4_dqpsk_diff_decoding.m*: Differential decoding and detection for $\pi/4 - DQPSK$

```
1  function [ a_cap,x,y ] = piBy4_dqpsk_Diff_decoding(w,z)
2  %Phase Mapper for pi/4-DQPSK modulation
3  % w - differentially coded I-channel bits at the receiver
4  % z - differentially coded Q-channel bits at the receiver
5  % a_cap - binary bit stream after differential decoding
6
7  if length(w)~=length(z), error('Length mismatch between w and z'); end
8
9  x = zeros(1,length(w)-1);y = zeros(1,length(w)-1);
10
11 for k=1:length(w)-1,
```

```
12      x(k) = w(k+1)*w(k) + z(k+1)*z(k);
13      y(k) = z(k+1)*w(k) - w(k+1)*z(k);
14  end
15  a_cap = zeros(1,2*length(x));
16  a_cap(1:2:end) = x > 0; %odd bits
17  a_cap(2:2:end) = y > 0; %even bits
18  end
```

Program 2.15: *piBy4_dqpsk_demod.m* : $\pi/4 - DQPSK$ demodulator

```
1   function [a_cap,x,y] = piBy4_dqpsk_demod(r,fc,OF)
2   %Function for differential coherent demodulation of pi/4-DQPSK
3   %r - received signal at the receiver front end
4   %fc - carrier frequency in Hertz
5   %OF - oversampling factor (multiples of fc) - at least 4 is better
6   %L - upsampling factor on the inphase and quadrature arms
7   %a_cap - detected binary stream
8   fs = OF*fc; %sampling frequency
9   L = 2*OF; %samples in 2Tb duration
10  t=0:1/fs:(length(r)-1)/fs;
11
12  w=r.*cos(2*pi*fc*t); %I arm
13  z=-r.*sin(2*pi*fc*t); %Q arm
14  w = conv(w,ones(1,L));%integrate for L (Tsym=2*Tb) duration
15  z = conv(z,ones(1,L));%integrate for L (Tsym=2*Tb) duration
16  w = w(L:L:end);%I arm - sample at every symbol instant Tsym
17  z = z(L:L:end);%Q arm - sample at every symbol instant Tsym
18
19  [a_cap,x,y] = piBy4_dqpsk_Diff_decoding(w,z);
20
21  doPlot=0; %To plot constellation at the receiver
22  if doPlot==1
23      figure; plot(w(1:200),z(1:200),'o');
24  end
```

Performance over AWGN channel

The wrapper code for simulating a complete communication chain with $\pi/4$-DQPSK modulator, AWGN channel and $\pi/4$-DQPSK demodulator is given next. The simulated waveforms at the transmitter are shown in Figure 2.20. The code also includes performance simulation of the $\pi/4$-DQPSK communication system over a range of E_b/N_0 values. The simulated performance curve is plotted in Figure 2.21 along with the theoretical bit error probability of differentially coherently demodulated $\pi/4$-DQPSK, which is given by

$$P_b = Q\left(\sqrt{\frac{4E_b}{N_0}}\,sin\frac{\pi}{4\sqrt{2}}\right)$$

$$= \frac{1}{2}\,erfc\left(\frac{1}{\sqrt{2}}\sqrt{\frac{4E_b}{N_0}}\,sin\frac{\pi}{4\sqrt{2}}\right) \tag{2.22}$$

Program 2.16: *piBy4_dqpsk_wfm_sim.m*: $\pi/4 - DQPSK$ - waveform and performance simulation

```matlab
%Performance of pi/4-DQPSK modulation (waveform simulation)
clear all;clc;
N=100000;%Number of symbols to transmit
EbN0dB = 0:2:14; % Eb/N0 range in dB for simulation
fc=800;%carrier frequency in Hertz
OF =16; %oversampling factor, sampling frequency will be fs=OF*fc

EbN0lin = 10.^(EbN0dB/10); %converting dB values to linear scale
BER = zeros(length(EbN0dB),1); %SER values for each Eb/N0

a = rand(N,1)>0.5; %random source symbols from 0's and 1's
[s,t] = piBy4_dqpsk_mod(a,fc,OF);%dqpsk modulation

for i=1:length(EbN0dB),
    Eb=OF*sum(abs(s).^2)/(length(s)); %OF is the over sampling factor
    N0= Eb/EbN0lin(i); %required noise spectral density from Eb/N0
    n = sqrt(N0/2)*(randn(1,length(s)));%computed noise
    r = s + n; %noisy received signal
    [a_cap] = piBy4_dqpsk_demod(r,fc,OF);%Differential coherent demod
    BER(i) = sum(a~=a_cap.')/N;%Symbol Error Rate Computation
end

x = sqrt(4*EbN0lin)*sin(pi/(4*sqrt(2)));
theoreticalBER = 0.5*erfc(x/sqrt(2));%Theoretical Bit Error Rate
figure;semilogy(EbN0dB,BER,'k*','LineWidth',1.5); %simulated BER
hold on; semilogy(EbN0dB,theoreticalBER,'r-','LineWidth',1.5);
title('Probability of Bit Error for \pi/4-DQPSK');
xlabel('E_b/N_0 (dB)');ylabel('Probability of Bit Error - P_b');
legend('Simulated', 'Theoretical');grid on;
```

2.8 Continuous Phase Modulation (CPM)

2.8.1 Motivation behind CPM

Modulation schemes like conventional PSK or FSK introduce abrupt phase discontinuities in the transmitted signal. Abrupt phase discontinuities produce sidelobes and could cause interference on the nearby channels. It is highly desirable that the chosen modulation scheme for a radio communication, should produce a spectrally efficient signal (minimized sidelobe effects) so that more users can be served in the allocated frequency band. Additionally, the phase discontinuities can also introduce undesirable frequency sidelobes when passed through power-efficient non-linear amplifiers (Class C).

In the case of conventional PSK or FSK modulation, the carrier phase abruptly resets to zero at the start of every transmitted symbol. If the carrier phase can be made continuous from one symbol to another symbol, the sidelobe levels can be reduced.

Offset QPSK (OQPSK) modulation, discussed in section 2.6, is a special case of QPSK modulation that limits the range of phase variations in the transmitted signal. Therefore, some improvement in the sidelobe levels can be achieved with OQPSK. Further improvement can be obtained with Continuous Phase Modulation

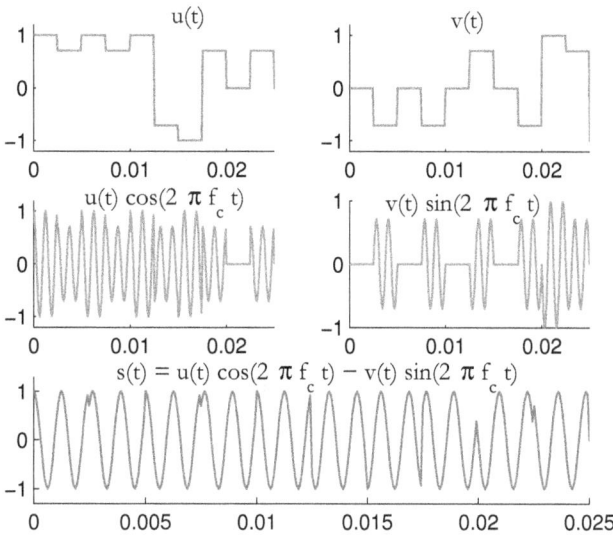

Fig. 2.20: Simulated $\pi/4$-DQPSK waveforms at the transmitter side

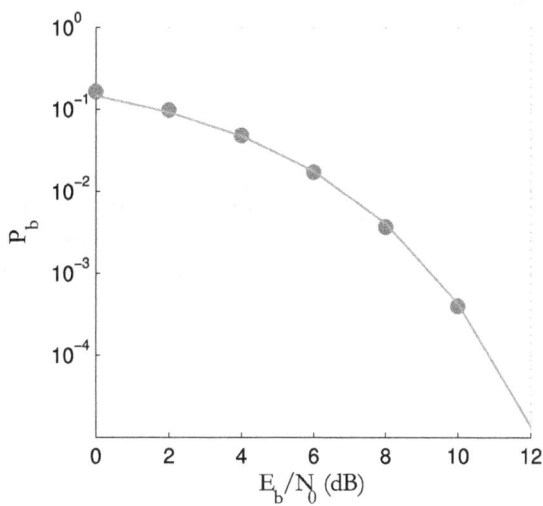

Fig. 2.21: Performance of differential coherently demodulated $\pi/4$-DQPSK

(CPM) schemes where the carrier phase smoothly varies from one symbol to another symbol in a continuous manner. It is important to note that OQPSK is not a continuous phase modulation scheme. It simply restricts the range of phase variations and does not make the phase variations continuous.

2.8.2 Continuous Phase Frequency Shift Keying (CPFSK) modulation

In CPM, the phase of the carrier is gradually changed from one symbol to another symbol. The instantaneous phase depends on the phase of previous symbols. Therefore CPM is a *modulation with memory*. The modulation and demodulation of CPM are more complicated by the fact that the phase of one symbol depends on the phase of the previous symbols. The phase trajectories of CPM modulated signals are best exploited by like MLSE (Maximum Likelihood Sequence Estimation) technique implemented by Viterbi algorithm. The demodulation techniques for CPM are more complex than the techniques used for memoryless modulation schemes like BPSK,QPSK,QAM,FSK etc.

Continuous Phase Frequency Shift Keying (CPFSK) is a variation of Frequency Shift Keying (FSK) technique that eliminates phase discontinuities. In conventional FSK modulation, the instantaneous carrier frequency is switched between one of the two frequencies. Due to the abrupt switching between the two frequencies, the phase is not continuous in conventional FSK. In binary CPFSK, the smooth phase continuity is ensured by defining the transmitter signal as

$$s(t) = \sqrt{\frac{2E_b}{T_b}} cos\left[2\pi f_c t + \theta(t)\right] \tag{2.23}$$

where, E_b is the average energy per bit, T_b is the bit period, f_c is the *base carrier frequency* and $\theta(t)$ is the *phase evolution* whose derivative gives rise to CPFSK's instantaneous angular frequency shift.

At any time instant, the phase evolution can be given as

$$\theta(t) = \theta(0) + \frac{\pi h}{T_b} \int_0^t b(t)dt \tag{2.24}$$

where $\theta(0)$ being the accumulated phase history till $t = 0$, the factor h is called *modulation index* which is a measure of frequency deviation ($h = 1$ corresponds to binary CPFSK and $h = 0.5$ corresponds to *Minimum Shift Keying* (MSK) modulation), $b(t) \in (\pm 1)$ is the waveform that represents the binary information sequence a: such that the value $b = +1$ represents $a = 0$ and the value $b = -1$ represents $a = 1$.

The following snippet of code simulates the binary CPFSK modulation ($h = 1$) using equations 2.23, 2.24 and the resulting waveforms are plotted in Figure 2.22. The phase trajectory plotted in Figure 2.22-(b) shows ambiguity in the phase transitions ($+\pi$ phase transition is equivalent to $-\pi$ transition due to modulo 2π arithmetic). Due to this, the receiver will not be able to exploit the phase information in the binary CPFSK.

Program 2.17: *cpfsk.m*: Binary CPFSK modulation

```
L=50;%oversampling factor
Tb=0.5;%bit period in seconds
fs=L/Tb;%sampling frequency in Hertz
fc=2/Tb; %carrier frequency
N = 8;%number of bits to transmit
h=0.5; %modulation index
b=2*(rand(N,1)>0.5)-1;%random information sequence in +1/-1 format
b=repmat(b,1,L).';%oversampling by L samples per bit
b=b(:).';%serialize
theta= pi*h/Tb*filter(1,[1 -1],b,0)/fs;%use FIR integrator filter
t=0:1/fs:Tb*N-1/fs; %time base
s = cos(2*pi*fc*t + theta); %CPFSK signal
subplot(3,1,1);plot(t,b);xlabel('t');ylabel('b(t)');
subplot(3,1,2);plot(t,theta);xlabel('t');ylabel('\theta(t)');
subplot(3,1,3);plot(t,s);xlabel('t');ylabel('s(t)');
```

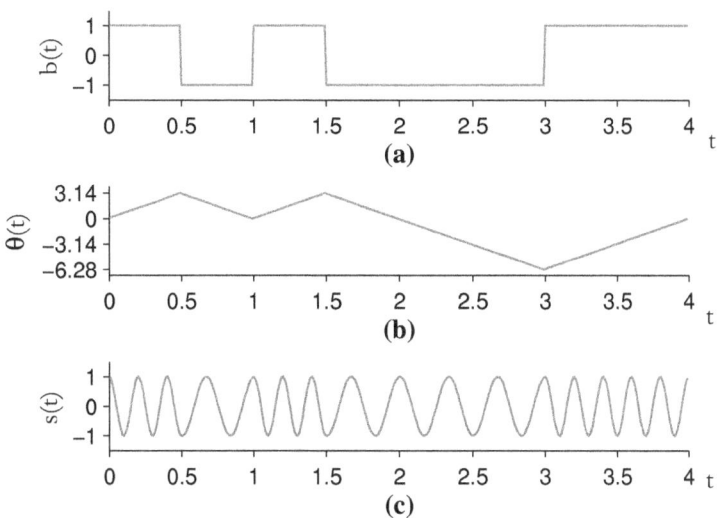

Fig. 2.22: (a) Information sequence, (b) phase evolution of CPFSK signal and (c) the CPFSK modulated signal

2.8.3 Minimum Shift Keying (MSK)

The Minimum Shift Keying is a true CPM modulation technique. MSK modulation provides all the desired qualities loved by the communication engineers - it provides constant envelope, a very compact spectrum compared to QPSK and OQPSK, and a good error rate performance.

MSK can be viewed as a special case of binary CPFSK where the modulation index h in the equation 2.24 is set to 0.5. The same snippet of code given for CPFSK simulation is executed with h set to 0.5 and the results are plotted in Figure 2.23.The phase trajectory of MSK in Figure 2.23-(b), reveals that each information bit leads to different phase transitions on modulo-2π. The receiver can exploit these phase transitions without any ambiguity and it can provide better error rate performance. This is the main motivation behind the MSK technique.

Without loss of generality, assuming $\theta(0) = 0$ and setting $h = 0.5$ in equation 2.24, the equation 2.23 is modified to generate an MSK signal as

$$s(t) = \sqrt{\frac{2E_b}{T_b}} cos \left[\theta(0) \pm \frac{\pi}{2T_b}t \right] cos\left(2\pi f_c t\right) - \sqrt{\frac{2E_b}{T_b}} sin \left[\theta(0) \pm \frac{\pi}{2T_b}t \right] sin\left(2\pi f_c t\right)$$
$$= s_I(t)cos\left(2\pi f_c t\right) - s_Q(t)sin\left(2\pi f_c t\right) \quad , 0 \leq t \leq 2T_b \tag{2.25}$$

where the inphase component $s_I(t)$ and the quadrature component $s_Q(t)$ can be re-written as

$$s_I(t) = \pm\sqrt{\frac{2E_b}{T_b}} cos \left[\frac{\pi}{2T_b}t \right] \quad , -T_b \leq t \leq T_b$$
$$s_Q(t) = \pm\sqrt{\frac{2E_b}{T_b}} sin \left[\frac{\pi}{2T_b}t \right] \quad , 0 \leq t \leq 2T_b \tag{2.26}$$

Therefore, rewriting equation 2.25, the MSK waveform is given by

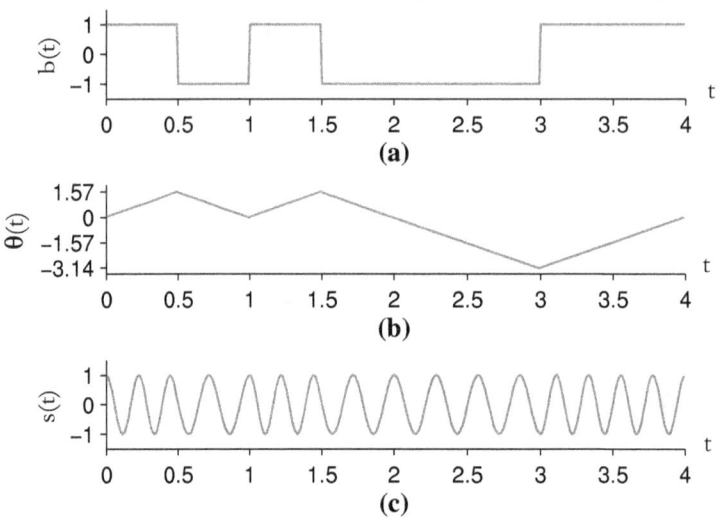

Fig. 2.23: (a) Information sequence, (b) phase evolution of MSK signal and (c) the MSK modulated signal

$$s(t) = \sqrt{\frac{2}{T_b}}\, a_I(t)\, cos\left[\frac{\pi}{2T_b}t\right]\, cos\,(2\pi f_c t) - \sqrt{\frac{2}{T_b}}\, a_Q(t)\, sin\left[\frac{\pi}{2T_b}t\right]\, sin\,(2\pi f_c t) \qquad (2.27)$$

where $a_I(t)$ and $a_Q(t)$ are random information sequences in the I-channel and Q-channel respectively.

Equivalence to OQPSK modulation

From equation 2.26, the inphase component $s_I(t)$ is interpreted as a half-cycle cosine function for the whole interval $(-T_b, T_b]$ and the quadrature component $s_Q(t)$ is interpreted as a half-cycle sine function for the interval $(0, 2T_b]$. Therefore, the half-cycle cosine and sine functions are offset from each other by T_b seconds. This offset relationship between the inphase and quadrature components is more similar to that of a OQPSK signal construct [6].

Figure 2.24 illustrates the similarities between the OQPSK and MSK signal construction. In OQPSK, the rectangular shaped inphase and quadrature components are offset by half symbol period ($T_{sym}/2 = T_b$ seconds). Whereas, in MSK modulation, the inphase and quadrature components are similarly offset by half symbol period ($T_{sym}/2 = T_b$ seconds) but they are additionally shaped by half-cycle cosine and sine functions. However, the half-cycle functions in the MSK are not simple reshaping waveforms. On the inphase arm, if the information bit sequence $a_I(t)$ is positive, then $s_I(t)$ follows $cos\{\pi t/(2T_b)\}$ and if $a_I(t)$ is negative, then $s_I(t)$ follows $-cos\{\pi t/(2T_b)\}$. Similarly, on the quadrature arm, if the information bit sequence $a_Q(t)$ is positive, then $s_Q(t)$ follows $sin\{\pi t/(2T_b)\}$ and if $s_Q(t)$ is negative, then $s_Q(t)$ follows $-sin\{\pi t/(2T_b)\}$

MSK modulator

Several forms of MSK signal generation/detection techniques exist and a good analysis can be found in reference [7]. As discussed above, one such form is viewing an MSK signal as a special form of OQPSK signal construct. A practical MSK modulator that is more similar to the OQPSK modulator structure is given in

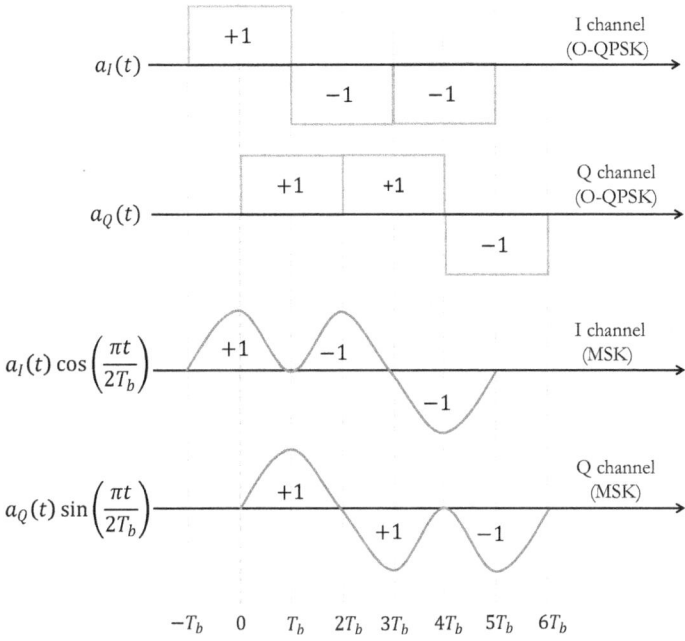

Fig. 2.24: Similarities between OQPSK and MSK waveforms

Figure 2.25 and the corresponding Matlab function (msk_mod) is also given next. Here, the pulse shaping functions on the inphase and quadrature arms are implemented using identical low-pass filters having the following impulse response.

$$h(t) = \begin{cases} sin\left(\frac{\pi}{2T_b}t\right) & ,0 \leq t \leq 2T_b \\ 0 & \text{, otherwise} \end{cases} \qquad (2.28)$$

Program 2.18: *msk_mod.m*: MSK modulator

```matlab
function [s,t,I,Q] = msk_mod(a,fc,OF)
%Function to modulate an incoming binary stream using MSK
%a - input binary data stream (0's and 1's) to modulate
%fc - carrier frequency in Hertz
%OF - oversampling factor
%s - MSK modulated signal with carrier
%t - time base for the carrier modulated signal
%I - baseband I channel waveform(no carrier)
%Q - baseband Q channel waveform(no carrier)
ak = 2*a-1; %NRZ encoding 0-> -1, 1->+1
ai = ak(1:2:end); aq = ak(2:2:end);%even and odd bit streams

L=2*OF;%represents one symbol duration Tsym=2xTb
%upsample and serialize bits streams in I and Q arms
ai = [ai zeros(length(ai),L-1)]; ai=ai.';ai=ai(:);
aq = [aq zeros(length(aq),L-1)]; aq=aq.';aq=aq(:);

ai = [ai(:) ; zeros(L/2,1)].'; %padding at end to equal length of Q
```

```
19  aq = [zeros(L/2,1); aq(:)].'; %delay by Tb (delay by L/2)
20
21  %construct Low-pass filter and filter the I/Q samples through it
22  Fs=OF*fc;Ts=1/Fs;Tb = OF*Ts;
23  t=0:Ts:2*Tb; h = sin(pi*t/(2*Tb));%LPF filter
24  I = filter(h,1,ai);%baseband I-channel
25  Q = filter(h,1,aq);%baseband Q-channel
26
27  t=(0:1:length(I)-1)*Ts; %for RF carrier
28  iChannel = I.*cos(2*pi*fc*t); qChannel = Q.*sin(2*pi*fc*t);
29  s = iChannel-qChannel; %Bandpass MSK modulated signal
30
31  doPlot=1; %set to 0 if you do not intend to see waveform plots
32  if doPlot==1,%Waveforms at the transmitter
33    figure;
34    subplot(3,1,1);hold on;plot(t,I,'--');plot(t,iChannel,'r');
35    xlabel('t'); ylabel('s_I(t)');xlim([-Tb,20*Tb]);
36    subplot(3,1,2);hold on;plot(t,Q,'--');plot(t,qChannel,'r');
37    xlabel('t'); ylabel('s_Q(t)');xlim([-Tb,20*Tb]);
38
39    subplot(3,1,3);plot(t,s,'k'); %MSK wfm zoomed to first few symbols
40    xlabel('t'); ylabel('s(t)');xlim([-Tb,20*Tb]);
41    %constellation plot at transmitter
42    figure;plot(I(40:end-20),Q(40:end-20));axis([-1.5 1.5 -1.5 1.5]);
43  end
44  end
```

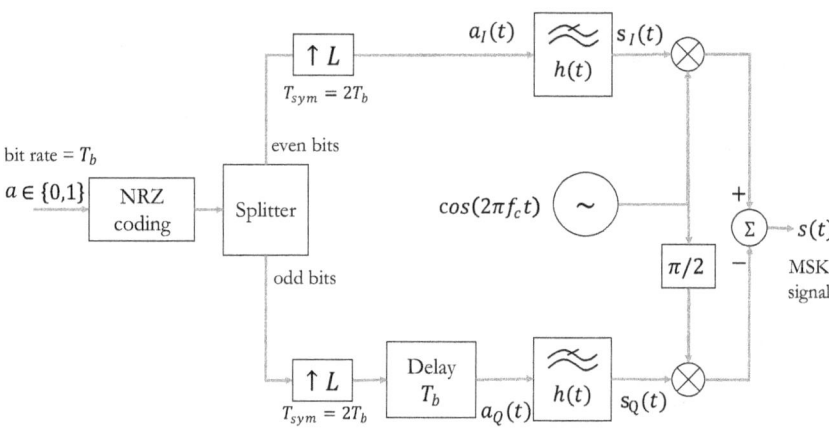

Fig. 2.25: A practical MSK modulator constructed according to the equivalence with OQPSK modulation

MSK demodulator

The corresponding demodulator is given in Figure 2.26 and the Matlab function for MSK demodulation (msk_demod) is also given next. Note that at the demodulator, the signal in the inphase and quadrature arms are multiplied by the absolute value of the half-cycle functions: $|cos\{\pi t/(2T_b)\}|$ and $|sin\{\pi t/(2T_b)\}|$, respectively. The rest of the demodulator is same as the OQPSK demodulator structure shown in Figure 2.15.

Program 2.19: *msk_demod.m*: MSK demodulator

```
1   function [a_cap] = msk_demod(r,N,fc,OF)
2   %Function to demodulate a bandpass MSK signal
3   %r - received signal at the receiver front end
4   %N - number of symbols transmitted
5   %fc - carrier frequency in Hertz
6   %OF - oversampling factor (multiples of fc) - at least 4 is better
7   %a_cap - detected binary stream
8
9   L = 2*OF; %samples in 2Tb duration
10  Fs=OF*fc;Ts=1/Fs;Tb = OF*Ts; %sampling frequency, durations
11  t=(-OF:1:length(r)-OF-1)/Fs; %time base
12
13  %cosine and sine functions for half-sinusoid shaping
14  x=abs(cos(pi*t/(2*Tb)));y=abs(sin(pi*t/(2*Tb)));
15
16  u=r.*x.*cos(2*pi*fc*t); %multiply I by half cosines and cos(2pifct)
17  v=-r.*y.*sin(2*pi*fc*t);%multiply Q by half sines and sin(2pifct)
18  iHat = conv(u,ones(1,L));%integrate for L (Tsym=2*Tb) duration
19  qHat = conv(v,ones(1,L));%integrate for L (Tsym=2*Tb) duration
20  iHat= iHat(L:L:end-L);%I- sample at the end of every symbol
21  qHat= qHat(L+L/2:L:end-L/2);%Q-sample from L+L/2th sample
22
23  a_cap = zeros(N,1);
24  a_cap(1:2:end) = iHat > 0; %thresholding - odd bits
25  a_cap(2:2:end) = qHat > 0; %thresholding - even bits
```

Performance simulation

The wrapper code for simulating a complete communication chain with the aforementioned MSK modulator, AWGN channel and the MSK demodulator is given next (*msk_wfm_sim.m*). The simulated timing waveforms at the transmitter are shown in Figure 2.27. The code also includes performance simulation of the MSK communication system over a range of E_b/N_0 values pertaining to the AWGN channel.

   MSK modulation can be interpreted as having two BPSK-like modulators on the I-channel and Q-channel arms and the only difference being the half-symbol shift with the half-cycle sinusoidal shaping for MSK. Similar argument applies for MSK demodulator as well. Hence the theoretical bit error probability of MSK is identical to that of conventional BPSK (see Figure 2.6 for reference), which is given by

$$P_b = \frac{1}{2}\, erfc\left(\sqrt{\frac{E_b}{N_0}}\right) \tag{2.29}$$

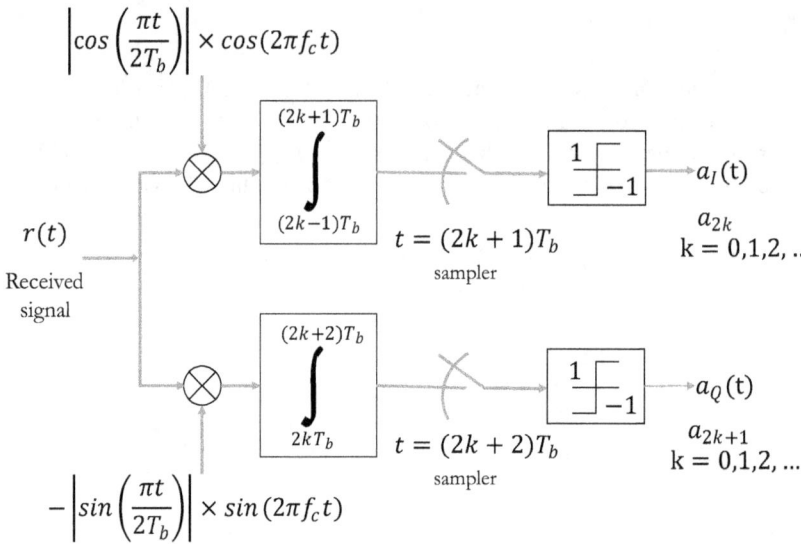

Fig. 2.26: MSK demodulator using the complex representation approach

Program 2.20: *msk_wfm_sim.m*: MSK demodulator

```
1  % Demonstration of Eb/N0 Vs SER for bandpass MSK (waveform sim)
2  clear all;clc;
3  N=100000;%Number of symbols to transmit
4  EbN0dB = -4:2:10; % Eb/N0 range in dB for simulation
5  fc=800;%carrier frequency in Hertz
6  OF =32; %oversampling factor, sampling frequency will be fs=OF*fc
7
8  EbN0lin = 10.^(EbN0dB/10); %converting dB values to linear scale
9  BER = zeros(length(EbN0dB),1); %SER values for each Eb/N0
10
11 a = rand(N,1)>0.5; %random symbols from 0's and 1's - input to QPSK
12 [s,t] = msk_mod(a,fc,OF);%MSK modulation
13
14 for i=1:length(EbN0dB),
15     Eb=OF*sum(abs(s).^2)/(length(s)); %energy per bit from samples
16     N0= Eb/EbN0lin(i); %required noise spectral density from Eb/N0
17     n = sqrt(N0/2)*(randn(1,length(s)));%computed noise
18     r = s + n;
19     %---------------Receiver--------------------
20     a_cap = msk_demod(r,N,fc,OF);
21     BER(i) = sum(a~=a_cap)/N;%Symbol Error Rate Computation
22 end
23 theoreticalBER = 0.5*erfc(sqrt(EbN0lin));%Theoretical BER
24
25 figure;semilogy(EbN0dB,BER,'k*','LineWidth',1.5); hold on;
26 semilogy(EbN0dB,theoreticalBER,'r-','LineWidth',1.5);
27 set(gca,'XLim',[-4 12]);set(gca,'YLim',[1E-6 1E0]);set(gca,'XTick',-4:2:12);
28 title('Probability of Bit Error for MSK modulation');
```

```
29  xlabel('E_b/N_0 (dB)');ylabel('Probability of Bit Error - P_b');
30  legend('Simulated', 'Theoretical');grid on;
```

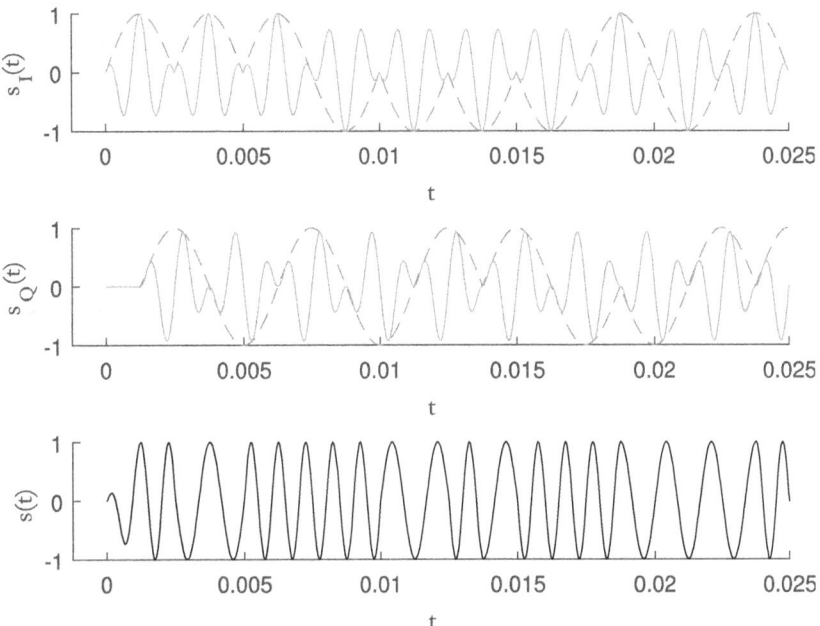

Fig. 2.27: Simulated MSK waveforms at the transmitter side

2.9 Investigating phase transition properties

The phase transition properties of the different variants of QPSK schemes and MSK, are easily investigated using constellation plots. The next code sample demonstrates how to plot the signal space constellations, for the various modulations used in the transmitter.

Typically, in practical applications, the baseband modulated waveforms are passed through a pulse shaping filter for combating the phenomenon of intersymbol interference (ISI). The goal is to plot the constellation plots of various pulse-shaped baseband waveforms of the QPSK, O-QPSK and $\pi/4$-DQPSK schemes. A variety of pulse shaping filters are available and raised cosine filter is specifically chosen for this demo.

The Raised Cosine (RC) pulse comes with an adjustable transition band roll-off parameter α, using which the decay of the transition band can be controlled. The RC pulse shaping function is expressed in frequency domain as

$$
P(f) = \begin{cases} T_{sym}, & |f| \leq \frac{1-\alpha}{2T_{sym}} \\ \frac{T_{sym}}{2}\left[1+cos\left(\frac{\pi T_{sym}}{\alpha}\left[|f|-\frac{1-\alpha}{2T_{sym}}\right]\right)\right], & \frac{1-\alpha}{2T_{sym}} \leq f \leq \frac{1+\alpha}{2T_{sym}} \\ 0, & |f| \geq \frac{1+\alpha}{2T_{sym}} \end{cases}
\tag{2.30}
$$

Equivalently, in time domain, the impulse response corresponds to

$$p(t) = \frac{sin\left(\frac{\pi t}{T_{sym}}\right)}{\frac{\pi t}{T_{sym}}} \frac{cos\left(\frac{\pi \alpha t}{T_{sym}}\right)}{1 - \left(\frac{2\alpha t}{T_{sym}}\right)^2} \quad where \; 0 \le \alpha \le 1 \tag{2.31}$$

A simple evaluation of the equation 2.31 produces singularities (undefined points) at $p(t = 0)$ and $p(t = \pm T_{sym}/(2\alpha))$. The value of the raised cosine pulse at these singularities can be obtained by applying L'Hospital's rule [8] and the values are

$$p(t = 0) = 1 \tag{2.32}$$

$$p\left(t = \pm \frac{T_{sym}}{2\alpha}\right) = \frac{\alpha}{2} sin\left(\frac{\pi}{2\alpha}\right) \tag{2.33}$$

Using the equations above, the raised cosine filter is implemented as a function:

Program 2.21: *raisedCosineFunction.m*: Implementing a Raised Cosine pulse shaper

```
function [p,t,filtDelay]=raisedCosineFunction(alpha,L,Nsym)
%Generatea raised cosine (RC) pulse
%alpha - roll-off factor,
%L - oversampling factor
%Nsym - filter span in symbols
%Returns the output pulse p(t) that spans the discrete-time
%base -Nsym:1/L:Nsym. Also returns the filter delay when the
%function is viewed as an FIR filter

Tsym=1; t=-(Nsym/2):1/L:(Nsym/2); % +/- discrete-time base
A = sin(pi*t/Tsym)./(pi*t/Tsym); B=cos(pi*alpha*t/Tsym);

%handle singularities at p(0) and p(t=+/-1/2a)
p = A.*B./(1-(2*alpha*t/Tsym).^2);
p(ceil(length(p)/2))=1; %p(0)=1 and p(0) occurs at center
temp=(alpha/2)*sin(pi/(2*alpha));
p(t==Tsym/(2*alpha))=temp; p(t==-Tsym/(2*alpha))=temp;
filtDelay = (length(p)-1)/2; %FIR filter delay = (N-1)/2
end
```

The function is tested with the following code. It generates a raised cosine pulse for the given symbol duration $T_{sym} = 1s$ and plots the time-domain view and the frequency response as shown in Figure 2.28. From the plot, it can be observed that the RC pulse falls off at the rate of $1/|t|^3$ as $t \to \infty$, which is a significant improvement when compared to the decay rate of a sinc pulse which is $1/|t|$. It satisfies Nyquist criterion for zero ISI - the pulse hits zero crossings at desired sampling instants. The transition bands in the frequency domain can be made gradual (by controlling α) when compared to that of a sinc pulse.

Program 2.22: *test_RCPulse.m*: Raised Cosine pulses and their manifestation in frequency domain

```
Tsym=1; %Symbol duration in seconds
L=10; % oversampling rate, each symbol contains L samples
Nsym = 80; %filter span in symbol durations
alphas=[0 0.3 0.5 1];%RC roll-off factors - valid range 0 to 1

Fs=L/Tsym;%sampling frequency
lineColors=['b','r','g','k','c']; i=1;legendString=cell(1,4);
```

```
8
9    for alpha=alphas %loop for various alpha values
10       [rcPulse,t]=raisedCosineFunction(alpha,L,Nsym); %RC Pulse
11       subplot(1,2,1); t=Tsym*t; %time base for symbol duration
12       plot(t,rcPulse,lineColors(i));hold on; %plot time domain view
13
14       %Compute FFT and plot double sided frequency domain view
15       NFFT=2^nextpow2(length(rcPulse)); %FFT length
16       vals=fftshift(fft(rcPulse,NFFT));
17       freqs=Fs*(-NFFT/2:NFFT/2-1)/NFFT;
18       subplot(1,2,2);
19       plot(freqs,abs(vals)/abs(vals(length(vals)/2+1)),lineColors(i));
20       hold on;legendString{i}=strcat('\alpha =',num2str(alpha) );i=i+1;
21    end
22    subplot(1,2,1);title('Raised Cosine pulse'); legend(legendString);
23    subplot(1,2,2);title('Frequency response');legend(legendString);
```

Now that we have constructed a function for raised cosine pulse shaping filter, the next step is to generate modulated waveforms (using QPSK, O-QPSK and $\pi/4$-DQPSK, MSK schemes), pass them through a raised cosine filter having a roll-off factor, say $\alpha = 0.3$ and finally plot the constellation.

Program 2.23: *constellations_plots.m*:Constellations of RC filtered waveforms of MSK and QPSK variants

```
1    clearvars; clc;
2    N=1000;%Number of symbols to transmit, keep it small and adequate
3    fc=10; L=8;%carrier frequency and oversampling factor
4
5    a = rand(N,1)>0.5; %random source symbols - 0's and 1's
6
7    %modulate the source symbols using QPSK,QPSK,pi/4-DQPSK and MSK
8    [s_qpsk,t_qpsk,I_qpsk,Q_qpsk]=qpsk_mod(a,fc,L);
9    [s_oqpsk,t_oqpsk,I_oqpsk,Q_oqpsk]=oqpsk_mod(a,fc,L);
10   [s_piBy4,t_piBy4,I_piBy4,Q_piBy4]=piBy4_dqpsk_mod(a,fc,L);
11   [s_msk,t_msk,I_msk,Q_msk] = msk_mod(a,fc,L);
12
13   %Pulse shape the modulated waveforms by convolving with RC filter
14   alpha = 0.3; Nsym = 10;% RC filter alpha and filter span in symbols
15   rcPulse = raisedCosineFunction(alpha,L,Nsym);%RC function
16   iRC_qpsk = conv(I_qpsk,rcPulse,'same');%RC shaped QPSK I channel
17   qRC_qpsk = conv(Q_qpsk,rcPulse,'same');%RC shaped QPSK Q channel
18   iRC_oqpsk = conv(I_oqpsk,rcPulse,'same');%RC shaped OQPSK I channel
19   qRC_oqpsk = conv(Q_oqpsk,rcPulse,'same');%RC shaped OQPSK Q channel
20   iRC_piBy4 = conv(I_piBy4,rcPulse,'same');%RC shpd pi/4-DQPSK I chan
21   qRC_piBy4 = conv(Q_piBy4,rcPulse,'same');%RC shpd pi/4-DQPSK Q chan
22
23   %Plot constellations
24   subplot(2,2,1);plot(iRC_qpsk,qRC_qpsk); %RC shaped QPSK
25   title('QPSK, RC \alpha=0.3');xlabel('I(t)');ylabel('Q(t)');
26   subplot(2,2,2);plot(iRC_oqpsk,qRC_oqpsk); %RC shaped O-QPSK
27   title('O-QPSK, RC \alpha=0.3');xlabel('I(t)');ylabel('Q(t)');
28   subplot(2,2,3);plot(iRC_piBy4,qRC_piBy4); %RC shaped pi/4 DQPSK
```

```
29   title('\pi/4-QPSK, RC \alpha=0.3');xlabel('I(t)');ylabel('Q(t)');
30   subplot(2,2,4);plot(I_msk(20:end-20),Q_msk(20:end-20)); %For MSK
31   title('MSK');xlabel('I(t)');ylabel('Q(t)');
```

Fig. 2.28: Raised Cosine pulse and its manifestation in frequency domain

The resulting simulated plot is shown in the Figure 2.29. From the resulting constellation plot, following conclusions can be reached.

- Conventional QPSK has 180° phase transitions and hence it requires linear amplifiers with high Q factor
- The phase transitions of Offset-QPSK are limited to 90° (the 180° phase transitions are eliminated)
- The signaling points for $\pi/4$-DQPSK is toggled between two sets of QPSK constellations that are shifted by 45° with respect to each other. Both the 90° and 180° phase transitions are absent in this constellation. Therefore, this scheme produces the lower envelope variations than the rest of the two QPSK schemes.
- MSK is a continuous phase modulation, therefore no abrupt phase transition occurs when a symbol changes as indicated by the smooth circumference in the constellation plot. Hence, a bandlimited MSK signal will not suffer any envelope variation, whereas, the rest of the QPSK schemes suffer varied levels of envelope variations, when they are bandlimited.

2.10 Power Spectral Density (PSD) plots

Power Spectral Density (PSD) is a measure of a signal's power intensity in the frequency domain. Comparison of PSD's of various modulation schemes plays a vital role in understanding their spectral characteristics and to make decisions on choosing the modulation scheme that satisfies the requirements of a given communication system. A discussion on PSD estimates is given in chapter 1 section 1.4.

The function plotWelchPSD, given in chapter chapter 1 section 1.4, is utilized here for plotting the PSDs of BPSK, QPSK and MSK modulated signals with carrier frequency of 800Hz and the resulting plot is given in Figure 2.30.

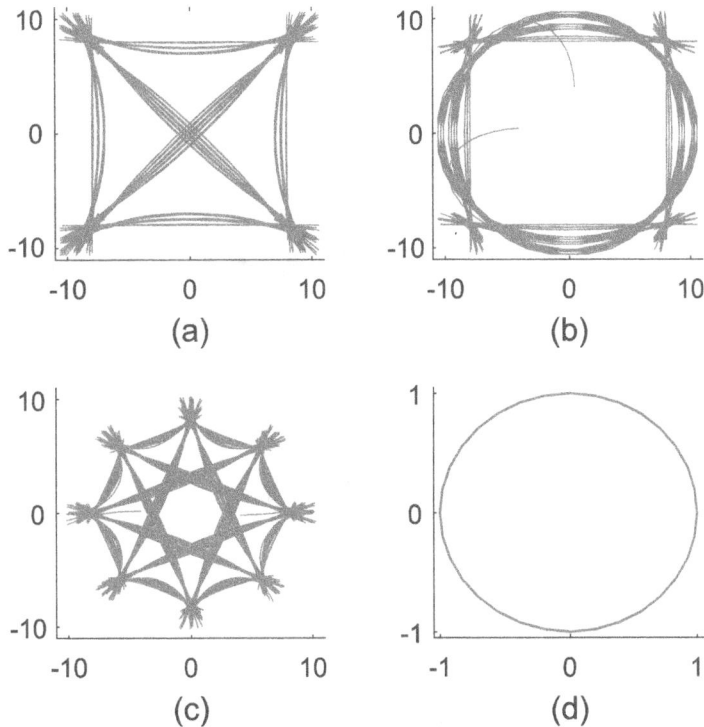

Fig. 2.29: Constellations plots for: (a) $\alpha = 0.3$ RC-filtered QPSK, (b) $\alpha = 0.3$ RC-filtered O-QPSK, (c) $\alpha = 0.3$ RC-filtered $\pi/4$ -DQPSK and (d) MSK

Program 2.24: *bpsk_qpsk_msk_psd.m*: Compute and plot PSD estimates of BPSK QPSK and MSK signals

```
%Compare PSDs of bandpass MSK QPSK BPSK signals
clear all;clc;
N=100000;%Number of symbols to transmit
Fc=800;OF =8;%carrier frequency and oversamping factor
Fs = Fc*OF;%sampling frequency

a = rand(N,1)>0.5; %random symbols to modulate

[s_bb,t]= bpsk_mod(a,OF); %BPSK modulation(waveform) - baseband
s_bpsk = s_bb.*cos(2*pi*Fc*t/Fs);%BPSK with carrier

s_qpsk = qpsk_mod(a,Fc,OF); %conventional QPSK
s_msk = msk_mod(a,Fc,OF);%MSK signal

%Compute and plot PSDs for each of the modulated versions
plotWelchPSD(s_bpsk,Fs,Fc,'b'); hold on;
plotWelchPSD(s_qpsk,Fs,Fc,'r');
plotWelchPSD(s_msk,Fs,Fc,'k');
legend('BPSK','QPSK','MSK');xlabel('f-f_c'),ylabel('PSD');
```

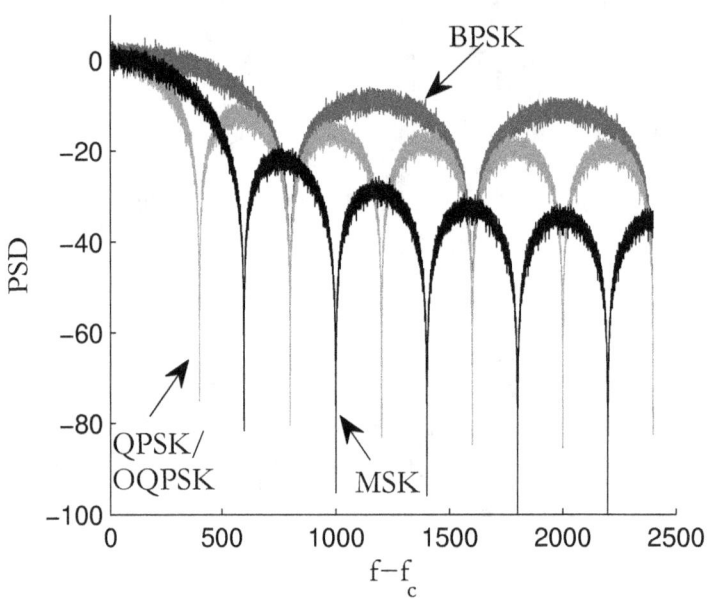

Fig. 2.30: PSD estimates for BPSK, QPSK and MSK signals

2.11 Gaussian Minimum Shift Keying (GMSK)

Minimum Shift Keying (MSK) is a special case of binary CPFSK with modulation index $h = 0.5$. It has features such as constant envelope, compact spectrum and good error rate performance. The fundamental problem with MSK is that the spectrum is not compact enough to satisfy the stringent requirements with respect to out-of-band radiation for technologies like GSM and DECT standard. These technologies have very high data rates approaching the RF channel bandwidth. A plot of MSK spectrum (Figure 2.30) will reveal that the sidelobes with significant energy, extend well beyond the transmission data rate. This is problematic, since it causes severe out-of-band interference in systems with closely spaced adjacent channels.

To satisfy such requirements, the MSK spectrum can be easily manipulated by using a pre-modulation low pass filter (LPF). The pre-modulation LPF should have the following properties and it is found that a Gaussian LPF will satisfy all of them [9].

- Sharp cut-off and narrow bandwidth - needed to suppress high frequency components.
- Lower overshoot in the impulse response - providing protection against excessive instantaneous frequency deviations.
- Preservation of filter output pulse area - thereby coherent detection can be applicable.

2.11.1 Pre-modulation Gaussian Low Pass Filter

Gaussian Minimum Shift Keying (GMSK) is a modified MSK modulation technique, where the spectrum of MSK is manipulated by passing the rectangular shaped information pulses through a Gaussian LPF *prior* to the frequency modulation of the carrier. A typical Gaussian LPF, used in GMSK modulation standards, is defined by the zero-mean Gaussian (bell-shaped) impulse response.

$$h(t) = \frac{1}{\sqrt{2\pi\sigma^2}} exp\left(\frac{-t^2}{2\sigma^2}\right) \quad, \sigma^2 = \frac{ln(2)}{(2\pi B)^2} \tag{2.34}$$

The parameter B is the 3-dB bandwidth of the LPF, which is determined from a parameter called BT_b as discussed next. If the input to the filter is an isolated unit rectangular pulse ($p(t) = 1, 0 \leq t \leq T_b$), the response of the filter will be [10]

$$g(t) = \frac{1}{2T_b}\left[Q\left(\frac{2\pi BT_b}{\sqrt{ln\ 2}}\left(\frac{t}{T_b} - 1\right)\right) - Q\left(\frac{2\pi BT_b}{\sqrt{ln\ 2}}\frac{t}{T_b}\right)\right] \tag{2.35}$$

where,

$$Q(x) = \int_x^\infty \frac{1}{\sqrt{2\pi}} exp\left(-\frac{y^2}{2}\right) dy, \quad -\infty \leq t \leq \infty$$

It is important to note the distinction between the two equations - 2.34 and 2.35. The equation for $h(t)$ defines the impulse response of the LPF, whereas the equation for $g(t)$, also called as *frequency pulse shaping function*, defines the LPF's output when the filter gets excited with a rectangular pulse. This distinction is captured in Figure 2.31.

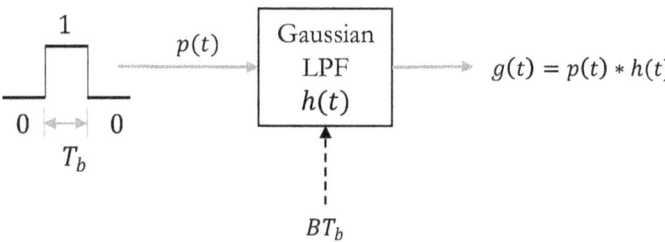

Fig. 2.31: Gaussian LPF: Relating $h(t)$ and $g(t)$

The aim of using GMSK modulation is to have a controlled MSK spectrum. Effectively, a variable parameter called BT_b, the product of 3-dB bandwidth of the LPF and the desired data-rate T_b, is often used by the designers to control the amount of spectrum efficiency required for the desired application. As a consequence, the 3-dB bandwidth of the aforementioned LPF is controlled by the BT_b design parameter. The range for the parameter BT_b is given as $0 < BT_b \leq \infty$. When $BT_b = \infty$, the impulse response $h(t)$ becomes a Dirac delta function $\delta(t)$, resulting in a transparent LPF and hence this configuration corresponds to MSK modulation.

The function to implement the Gaussian LPF's impulse response (equation 2.34), is given next. The Gaussian impulse response is of infinite duration and hence in digital implementations it has to be defined for a finite interval, as dictated by the function argument k in the code shown next. For example, in GSM standard, BT_b is chosen as 0.3 and the time truncation is done to three bit-intervals $k = 3$.

It is also necessary to normalize the filter coefficients of the computed LPF as

$$h[n] = \frac{h[n]}{\sum\limits_{i=0}^{N-1} h[i]}, \quad n = 0, 1, ..., N-1 \tag{2.36}$$

Program 2.25: *gaussianLPF.m*: Generate impulse response of the Gaussian LPF

```
1   function [h,t] = gaussianLPF(BT,Tb,L,k)
2   %Function to generate impulse response of a Gaussian low pass filter
3   %BT - BT product - Bandwidth x bit period
4   %Tb - bit period
5   %L - oversampling factor (number of samples per bit)
6   %k  - span length of the pulse (bit interval).
7   %h - impulse response of the Gaussian pulse
8   %t - generated time base
9   B = BT/Tb;%bandwidth of the filter
10  t=-k*Tb:Tb/L:k*Tb; %truncated time limits for the filter
11  h = sqrt(2*pi*B^2/(log(2)))*exp(-t.^2*2*pi^2*B^2/(log(2)));
12  h=h/sum(h);
```

Based on the gaussianLPF function , given above, we can compute and plot the impulse response $h(t)$ and the response to an isolated unit rectangular pulse - $g(t)$. The resulting plot is shown in Figure 2.32.

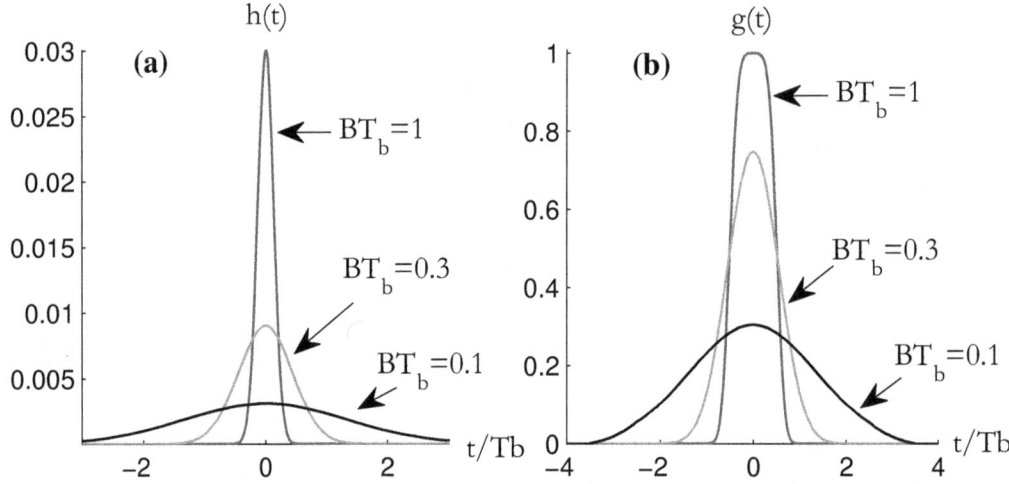

Fig. 2.32: Gaussian LPF:(a) impulse response and (b) response to isolated rectangular pulse

2.11.2 *Quadrature implementation of GMSK modulator*

Different implementations of a GMSK transmitter are possible. For conventional designs based on CPM representation, refer [10]. Quadrature design is another implementation for GMSK modulator that can be easily realized in software. The quadrature modulator for GMSK, shown in Figure 2.33, is readily obtained from Continuous Phase Modulation representation as

$$s(t) = cos\left(2\pi f_c t + \frac{\pi h}{T_b}\int_{-\infty}^{t} b(\tau)d\tau\right)$$

$$= cos\left(2\pi f_c t\right)cos\left[\frac{\pi h}{T_b}\int_{-\infty}^{t} b(\tau)d\tau\right] - sin\left(2\pi f_c t\right)sin\left[\frac{\pi h}{T_b}\int_{-\infty}^{t} b(\tau)d\tau\right]$$

$$= cos\left(2\pi f_c t\right)cos\left[\phi(t)\right] - sin\left(2\pi f_c t\right)sin\left[\phi(t)\right]$$

$$= I(t)cos\left(2\pi f_c t\right) - Q(t)sin\left(2\pi f_c t\right) \tag{2.37}$$

where, $b(t)$ is a Gaussian filtered NRZ data waveform sequence defined as

$$b(t) = \sum_{n=-\infty}^{\infty} a_n g(t - nT_b) \tag{2.38}$$

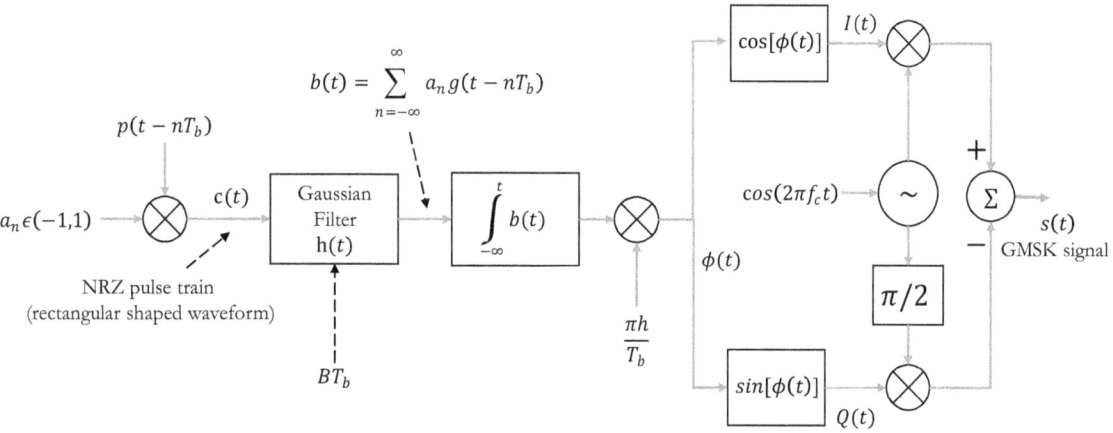

Fig. 2.33: Quadrature implementation of GMSK modulator

It can be readily seen that the generated GMSK signal depends on two parameters:

- BT_b - the product of 3-dB LPF bandwidth and the bit-period
- h - the *modulation index* defined as the ratio of peak-to-peak frequency deviation and the bit rate

$$h = \frac{\text{peak-to-peak frequency deviation}}{bit\,rate} = 2\Delta f \times T_b \tag{2.39}$$

GSM standard uses $BT_b = 0.3$ and $h = 0.5$ for generating the GMSK signal at the transmitter.

The function gmsk_mod implements the quadrature modulator (for $h = 0.5$). The simulated waveforms at various points in the modulator are given in Figure 2.34.

Program 2.26: *gmsk_mod.m*: Quadrature implementation of GMSK modulator

```
function [s,t,s_complex] = gmsk_mod(a,Fc,L,BT)
%Function to modulate a binary stream using GMSK modulation
%a - input binary data stream (0's and 1's) to modulate
%Fc - RF carrier frequency in Hertz
%L - oversampling factor
%BT - BT product (bandwidth x bit period) for GMSK
%s - GMSK modulated signal with carrier
```

```matlab
 8  %t - time base for the carrier modulated signal
 9  %s_complex - baseband GMSK signal (I+jQ)
10  Fs = L*Fc; Ts=1/Fs;Tb = L*Ts; %derived waveform timing parameters
11  a=a(:); %serialize input stream
12  ck = 2*a-1;%NRZ format
13  ct = kron(ck,ones(L,1));%Convert to waveform using oversampling
14
15  k=1;%truncation length for Gaussian LPF
16  ht = gaussianLPF(BT,Tb,L,k); %Gaussian LPF with BT=0.25
17  bt = conv(ht,ct,'full'); %convolve with Gaussian LPF
18  bt = bt/max(abs(bt));%normalize the output of Gaussian LPF to +/-1
19  phi = filter(1,[1,-1],bt*Ts);
20  phi = phi *0.5*pi/Tb; %h=0.5 - %integrate to get phase information
21
22  I = cos(phi);
23  Q = sin(phi); %cross-correlated baseband I/Q signals
24  s_complex = I - 1i*Q; %complex baseband representation
25
26  t=((0:1:length(I)-1)*Ts).'; %for RF carrier
27  iChannel = I.*cos(2*pi*Fc*t);
28  qChannel = Q.*sin(2*pi*Fc*t);
29  s =(iChannel - qChannel).'; %real signal - with RF carrier
30
31  doPlot=1;
32  if doPlot==1, figure;
33  subplot(2,4,1);
34  plot((0:1:length(ct)-1)*Ts,ct);title('c(t)');
35  subplot(2,4,2);
36  plot(-k*Tb:Ts:k*Tb,ht);title(['h(t)-BT_b=',num2str(BT)]);
37  subplot(2,4,5);
38  plot((0:1:length(bt)-1)*Ts,bt);title('b(t)');
39  subplot(2,4,6);
40  plot((0:1:length(phi)-1)*Ts,phi);title('\phi(t)');
41  subplot(2,4,3);
42  plot(t,I,'--');hold on;plot(t,iChannel,'r');
43  xlim([0,10*Tb]);title('I(t)cos(2 \pi f_c t)');
44  subplot(2,4,4);
45  plot(t,Q,'--');hold on;plot(t,qChannel,'r');
46  xlim([0,10*Tb]);title('Q(t)sin(2 \pi f_c t)');
47  subplot(2,4,7);
48  plot(t,s);title('s(t)');xlim([0,10*Tb]);
49  subplot(2,4,8); plot(I,Q);title('constellation');
50  end
51  end
```

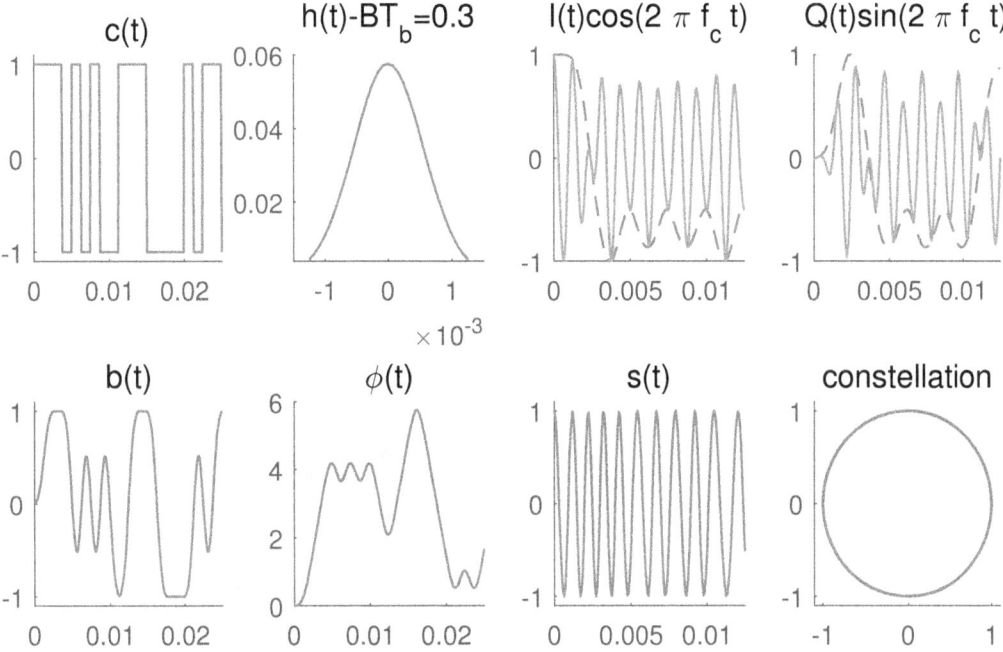

Fig. 2.34: Baseband waveforms at various points of the GMSK modulator

2.11.3 GMSK spectra

The PSD of GMSK signal is computed using the windowed Welch spectrum estimation method as described in chapter chapter 1 section 1.4. The PSDs of the GMSK signal for various BT_b values of the Gaussian LPF, are shown in Figure 2.35. For relatively small values of BT_b products (≤ 0.3), the truncation of the LPF co-efficients leads to spectrum floor. This result matches the documented observation in [11].

Program 2.27: *gmsk_psd.m*: PSD of GMSK using Welch spectrum estimation

```
1   % PSD of GMSK signals with various BT products
2   clearvars; clc;
3   N=10000;%Number of symbols to transmit
4   Fc=800;%carrier frequency in Hertz
5   L =16; %oversampling factor,use L= Fs/Fc, where Fs >> 2xFc
6   Fs = L*Fc;
7
8   a = rand(N,1)>0.5; %random symbols input to modulator
9   s1= gmsk_mod(a,Fc,L,0.3); %BT_b=0.3
10  s2 = gmsk_mod(a,Fc,L,0.5); %BT_b=0.5
11  s3 = gmsk_mod(a,Fc,L,0.7); %BT_b=0.7
12  s4 = gmsk_mod(a,Fc,L,10000); %BT_b=10000 (very large value-MSK)
13
14  %see section - 'Power Spectral Density plots' for definition
15  figure;
16  plotWelchPSD(s1,Fs,Fc,'r'); hold on;
17  plotWelchPSD(s2,Fs,Fc,'b');
```

```
18  plotWelchPSD(s3,Fs,Fc,'m');
19  plotWelchPSD(s4,Fs,Fc,'k');
20  xlabel('f-f_c'); ylabel('PSD (dB/Hz)');
21  legend('BT_b=0.3','BT_b=0.5','BT_b=1','BT_b=\infty');
```

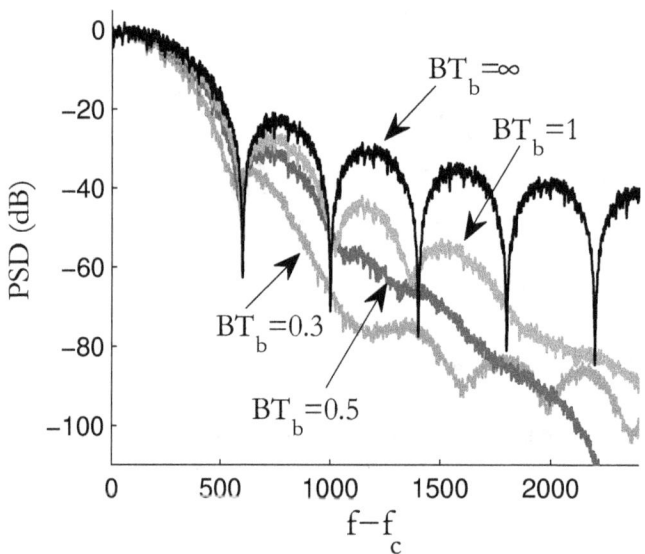

Fig. 2.35: GMSK spectrum using truncated impulse response (k=1 symbol)

2.11.4 GMSK demodulator

There exists numerous methods to demodulate a GMSK signal. A good survey of them can be found in the reference [11] and [12]. A cross-coupled IQ coherent demodulator, shown in Figure 2.36, is studied here for the simulation. The received signal, $r(t)$, is first converted to an intermediate frequency (IF) signal using a band pass filter (not shown in the figure). With the help of recovered carrier, the IF signal is translated to baseband using the IQ implementation. The LPF filters in the inphase and quadrature arms, are simply used to remove the $2f_{if}$ frequency components that result from the down conversion multipliers. The baseband signals $I(t)$ and $Q(t)$ can be used by the decision algorithm (MLSE, parallel/serial type MSK detection etc..,) for data detection.

From the modulator structure shown in Figure 2.33, it can be readily seen that the baseband inphase and quadrature components are correlated as

$$I(t) = cos[\phi(t)]$$
$$Q(t) = sin[\phi(t)] \tag{2.40}$$

At this stage, it is tempting to deduce the phase information $\phi(t)$ directly from the baseband quadrature components as

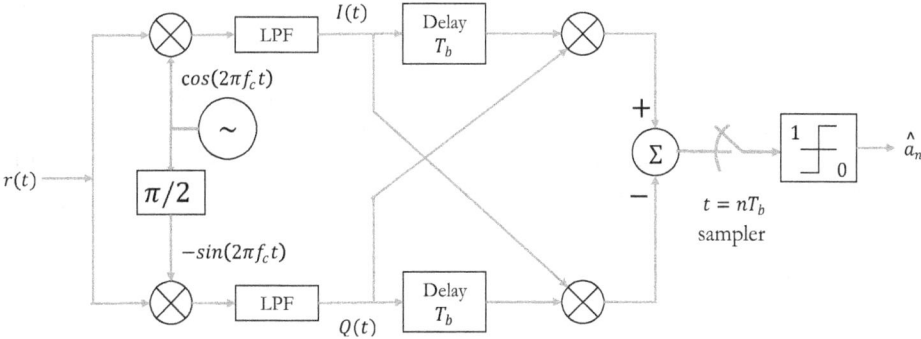

Fig. 2.36: Cross-coupled baseband IQ detector for GMSK demodulation

$$\phi(t) = tan^{-1}\left(\frac{Q(t)}{I(t)}\right) = ARCTAN2\left(\frac{Q(t)}{I(t)}\right) \tag{2.41}$$

Though, the above equation is mathematically correct, applying this direct approach for the detection at the receiver, is not a good solution, especially when the received signal is mixed with noise. Due to the addition of noise, any slight deviation in the detected instantaneous phase, leads to error propagation. Since the GMSK is a continuous phase modulation scheme, the correct method is to find the phase difference over a single bit interval and then build a decision logic using the phase difference.

The phase difference over one bit interval (T_b) is given as

$$\Delta\phi(t) = \phi(t) - \phi(t - T_b) \tag{2.42}$$

Taking $sin()$ on both sides and applying trigonometric identity for $sin(a - b)$,

$$\begin{aligned} sin[\Delta\phi(t)] &= sin[\phi(t) - \phi(t - T_b)] \\ &= sin[\phi(t)]cos[\phi(t - T_b)] - cos[\phi(t)]sin[\phi(t - T_b)] \\ &= Q(t)I(t - T_b) - I(t)Q(t - T_b) \end{aligned} \tag{2.43}$$

The data bits are decided by applying hard decision on the sign of the above cross-coupled term (now you can relate how the demodulator in Figure 2.36 got the cross-coupled structure).

The following function gmsk_demod implements the cross-coupled detector using equation 2.43. It operates on the baseband received signal represented in complex IQ form.

Program 2.28: *gmsk_demod.m*: Cross-coupled quadrature implementation of a GMSK demodulator

```
function a_cap = gmsk_demod(r,L)
%Function to demodulate a baseband GMSK signal
%r - received signal at receiver front end (complex form - I+jQ)
%L - oversampling factor
%a_cap - detected binary stream
I=real(r); Q = -imag(r); %I/Q streams
z1 = Q.*[zeros(L,1); I(1:length(I)-L)];
z2 = I.*[zeros(L,1); Q(1:length(I)-L)];
z = z1 - z2;
a_cap = z(2*L:L:end-L)>0;%sampling and hard decision
%sampling indices depends on the truncation length (k) of Gaussian LPF defined in
    the modulator
```

2.11.5 Performance

The error rate performance of the baseband IQ modulator and cross-coupled demodulator over an AWGN channel is simulated and the results are plotted in Figure 2.37. Further improvements are possible when other techniques like MLSE (Viterbi algorithm), equalization etc are applied for detection [12]. Describing all the possible methods for demodulation is beyond the scope of this text.

Program 2.29: *gmsk_wfm_sim.m*: Performance simulation of baseband GMSK over AWGN channel

```
1   % Demonstration of Eb/N0 Vs SER for baseband GMSK modulation scheme
2   clearvars ;clc;
3   N=100000;%Number of symbols to transmit
4   EbN0dB = 0:2:18; % Eb/N0 range in dB for simulation
5   Fc = 800; % Carrier frequency in Hz (must be < fs/2 and > fg)
6   BT= 0.3; %Gaussian LPF's BT product
7   L = 16; %oversampling factor
8
9   EbN0lin = 10.^(EbN0dB/10); %converting dB values to linear scale
10  BER = zeros(length(EbN0dB),1); %SER values for each Eb/N0
11
12  %-----------------Transmitter--------------------
13  a = rand(N,1)>0.5; %random symbols for modulation
14  [s,t,s_complex] = gmsk_mod(a,Fc,L,BT);%GMSK modulation
15
16  for i=1:length(EbN0dB),
17      Eb= sum(abs(s_complex).^2)/(length(s_complex)); %compute Energy
18      N0= Eb/EbN0lin(i); %required noise spectral density from Eb/N0
19      n = sqrt(N0/2)*(randn(size(s_complex))+1i*randn(size(s_complex)));
20      r = s_complex + n ; %noise added baseband GMSK signal
21      %--------------Receiver--------------------
22      a_cap = gmsk_demod(r,L);%Baseband GMSK demodulation
23      BER(i) = sum(a~=a_cap)/N;%Bit Error Rate Computation
24  end
25  figure;%Plot performance curves
26  semilogy(EbN0dB,BER,'g*-','LineWidth',1.5); hold on;%simulated BER
27  title('Probability of Bit Error for GMSK modulation');
28  xlabel('E_b/N_0 (dB)');ylabel('Probability of Bit Error - P_b');
```

2.12 Frequency Shift Keying (FSK)

FSK is a frequency modulation scheme in which the digital signal is modulated as discrete variations in the frequency of the carrier signal. The simplest FSK configuration, called *Binary FSK* (BFSK), uses two discrete frequencies to represent the binary data, one frequency for representing binary 0 and another frequency for representing binary 1. When extended to more frequencies it is called *M-ary FSK* or simply MFSK. Waveform simulation of BFSK modulator-demodulator, is discussed in this section. For simulating MFSK modulation-demodulation refer section 3.6 in chapter 3. Performance simulation of MFSK transmission is discussed in chapter 4.

Fig. 2.37: BER performance of GMSK IQ mod-demod (with Gaussian LPF response truncated to k=1 symbol)

2.12.1 Binary-FSK (BFSK)

In binary-FSK, the binary data is transmitted using two waveforms of frequencies f_1 and f_2 - that are slightly offset by Δf from the center of the carrier frequency f_c.

$$f_1 = f_c + \frac{\Delta f}{2}$$
$$f_2 = f_c - \frac{\Delta f}{2} \tag{2.44}$$

Sometimes, the frequency separation Δf is normalized to the bit-rate T_b using the *modulation index* as

$$h = \Delta f T_b \tag{2.45}$$

The BFSK waveform is generated as

$$S_{BFSK}(t) = \begin{cases} A\cos\left[2\pi f_1 t + \phi_1\right] = A\cos\left[2\pi\left(f_c + \frac{\Delta f}{2}\right)t + \phi_1\right], & 0 \le t \le T_b, \ for \ binary \ 1 \\ A\cos\left[2\pi f_2 t + \phi_2\right] = A\cos\left[2\pi\left(f_c - \frac{\Delta f}{2}\right)t + \phi_2\right], & 0 \le t \le T_b, \ for \ binary \ 0 \end{cases} \tag{2.46}$$

Here, ϕ_1 and ϕ_2 are the initial phases of the carrier at $t = 0$. If $\phi_1 = \phi_2$, then the modulation is called *coherent* BFSK, otherwise it is called *non-coherent* BFSK. From equations 2.45 and 2.46, the binary FSK scheme is represented as

$$S_{BFSK}(t) = \begin{cases} Acos\left[\left(2\pi f_c + \frac{\pi h}{T_b}\right)t + \phi_1\right], & 0 \leq t \leq T_b, \ for \ binary \ 1 \\ Acos\left[\left(2\pi f_c - \frac{\pi h}{T_b}\right)t + \phi_2\right], & 0 \leq t \leq T_b, \ for \ binary \ 0 \end{cases} \tag{2.47}$$

The basis functions of the BFSK can be made orthogonal in signal space by the proper selection of the modulation index h or the frequency separation Δf.

2.12.2 Orthogonality condition for non-coherent BFSK detection

Often, the frequencies f_1 and f_2 are chosen to be orthogonal. The choice of non-coherent or coherent FSK generation affects the orthogonality between the chosen frequencies. The orthogonality condition for non-coherent FSK ($\phi_1 \neq \phi_2$) is determined first and the results can be used to find the condition for coherent FSK ($\phi_1 = \phi_2 = \phi$).

The condition for orthogonality can be obtained by finding the correlation between the two waveforms with respect to the frequency shift Δf between the two frequencies or with respect to the modulation index h. The two frequencies are said to be orthogonal at places where the cross-correlation function equates to zero.

$$R(f_1, f_2) = R(\Delta f) = \int_0^{T_b} cos\,(2\pi f_1 t + \phi_1)\,cos\,(2\pi f_2 t + \phi_2)\,dt = 0 \tag{2.48}$$

Applying trigonometric identities, the correlation function can be written as

$$\frac{1}{2}\int_0^{T_b} cos\,(2\pi[f_1 + f_2]t + [\phi_1 + \phi_2])\,dt + \frac{1}{2}\int_0^{T_b} cos\,(2\pi[f_1 - f_2]t + [\phi_1 - \phi_2])\,dt = 0$$

$$\int_0^{T_b}\left[cos\,(2\pi[f_1 + f_2]t)\,cos(\phi_1 + \phi_2) - sin\,(2\pi[f_1 + f_2]t)\,sin(\phi_1 + \phi_2)\right]dt$$

$$+ \int_0^{T_b}\left[cos\,(2\pi[f_1 - f_2]t)\,cos\,(\phi_1 - \phi_2) - sin\,(2\pi[f_1 - f_2]t)\,sin\,(\phi_1 - \phi_2)\right]dt = 0 \tag{2.49}$$

Further expanding the terms and applying the limits after integration

$$cos(\phi_1 + \phi_2)\frac{sin\,(2\pi[f_1 + f_2]T_b)}{2\pi[f_1 + f_2]} + sin(\phi_1 + \phi_2)\left[\frac{cos\,(2\pi[f_1 + f_2]T_b) - 1}{2\pi[f_1 + f_2]}\right]$$

$$+cos\,(\phi_1 - \phi_2)\frac{sin\,(2\pi[f_1 - f_2]T_b)}{2\pi[f_1 - f_2]} + sin\,(\phi_1 - \phi_2)\left[\frac{cos\,(2\pi[f_1 - f_2]T_b) - 1}{2\pi[f_1 - f_2]}\right] = 0 \tag{2.50}$$

The above equation can be satisfied, if and only if, for some positive integers m and n,

$$2\pi[f_1 + f_2]T_b = 2m\pi \quad, m = 1, 2, 3, ...$$
$$2\pi[f_1 - f_2]T_b = 2n\pi \quad, n = 1, 2, 3, ... \tag{2.51}$$

This leads to the following solution.

$$f_1 = \frac{m+n}{2T_b}, \quad f_2 = \frac{m-n}{2T_b} \tag{2.52}$$

Therefore, the frequency separation for non-coherent case ($\phi_1 \neq \phi_2$) is given by

$$\Delta f = f_1 - f_2 = \frac{n}{T_b} \quad , n = 1, 2, 3, \ldots \tag{2.53}$$

Hence the minimum frequency separation for non-coherent FSK is obtained when $n = 1$. When expressed in terms of modulation index, using equation 2.45), the minimum frequency separation for obtaining orthogonal tones in the case of non-coherent FSK, occurs when $h = 1$.

$$(\Delta f)_{min} = \frac{1}{T_b} \Rightarrow h_{min} = 1 \tag{2.54}$$

2.12.3 Orthogonality condition for coherent BFSK

Noting that for coherent FSK generation, the initial phases are same for all frequency waveforms (i.e, $\phi_1 = \phi_2 = \phi$), the equation 2.50 shrinks to

$$cos(2\phi)\frac{sin\left(2\pi[f_1+f_2]T_b\right)}{2\pi[f_1+f_2]} + sin(2\phi)\left[\frac{cos\left(2\pi[f_1+f_2]T_b\right)-1}{2\pi[f_1+f_2]}\right] + \frac{sin\left(2\pi[f_1-f_2]T_b\right)}{2\pi[f_1-f_2]} = 0 \tag{2.55}$$

The above equation can be satisfied, if and only if, for some positive integers m and n,

$$2\pi[f_1+f_2]T_b = 2m\pi \quad , m = 1, 2, 3, \ldots$$
$$2\pi[f_1-f_2]T_b = n\pi \quad , n = 1, 2, 3, \ldots \tag{2.56}$$

This leads to

$$f_1 = \frac{2m+n}{4T_b}, \quad f_2 = \frac{2m-n}{4T_b}$$

Therefore, the frequency separation for coherent FSK ($\phi_1 = \phi_2$) is given by

$$\Delta f = f_1 - f_2 = \frac{n}{2T_b} \quad , n = 1, 2, 3, \ldots \tag{2.57}$$

Hence the minimum frequency separation for coherently detected FSK is obtained when $n = 1$. When expressed in terms of modulation index, using equation 2.45, the minimum frequency separation required for obtaining orthogonal tones for the case of coherent FSK, occurs when $h = 0.5$, (which corresponds to *Minimum Shift Keying* (MSK) should the phase change be continuous).

$$(\Delta f)_{min} = \frac{1}{2T_b} \Rightarrow h_{min} = 0.5 \tag{2.58}$$

For any positive integer n, when the frequency separation between the two tones is chosen as n/T_b, the phase of the coherent FSK is guaranteed to be continuous across bit transitions. But for coherent FSK, the minimum frequency separation for orthogonality is provided by $n = 0.5$ which will not guarantee a continuous phase across bit transitions. Minimum Shift Keying (MSK) is a particular form of coherent FSK that not only guarantees minimum frequency separation but also provides continuous phase at the bit transitions. Refer section 2.8.3 on MSK for more details.

It is evident from equations 2.54 and 2.58, while maintaining the orthogonality between the carrier tones, given the same bit rate T_b, coherent FSK occupies less bandwidth when compared to the non-coherently detected FSK. Thus, coherently detected FSK is more *bandwidth efficient* when compared to noncoherently detected FSK.

Table 2.2, captures the difference between coherent and non-coherent BFSK. The same concepts can be extended to M-ary FSK (MFSK).

Table 2.2: Summary of coherent and non-coherent BFSK schemes

	Non-coherent BFSK	Coherent BFSK
Initial phases during waveform generation	$\phi_1 \neq \phi_2$	$\phi_1 = \phi_2$
Orthogonality condition 1	f_1, f_2 must be integer multiples of $1/(2T_b)$	f_1, f_2 must be integer multiples of $1/(4T_b)$
Orthogonality condition 2	The difference $\Delta f = f_1 - f_2$ must be an integer multiple of $1/T_b$	The difference $\Delta f = f_1 - f_2$ must be an integer multiple of $1/(2T_b)$
Minimum frequency separation	$1/T_b$	$1/(2T_b)$
Modulation index for minimum frequency separation	$h = 1$	$h = 0.5$ (MSK)
Demodulation	can only be non-coherently demodulated	demodulation could be coherent or non-coherent
Performance in AWGN when coherently demodulated	cannot apply coherent demodulation	$Q\left(\sqrt{E_b/N_0}\right)$
Performance in AWGN when non-coherently demodulated	$0.5\, e^{-E_b/(2N_0)}$	$0.5\, e^{-E_b/(2N_0)}$

2.12.4 Modulator

The discrete time equivalent model for generating orthogonal BFSK signal is shown in Figure 2.38. Here, the discrete time sampling frequency (f_s) is chosen high enough when compared to the carrier frequency (f_c). The number of discrete samples in a bit-period L is chosen such that each bit-period is represented by sufficient number of samples to accommodate a few cycles of the frequencies of f_1 and f_2. The frequency separation Δf between f_1 and f_2 is computed from equation 2.45.

The function bfsk_mod implements the discrete-time equivalent model for BFSK generation, shown in Figure 2.38. It supports generating both coherent and noncoherent forms of BFSK signal. If the coherent BFSK is chosen, the function returns the generated initial random phase ϕ, in addition to the generated BFSK signal. This phase information is crucial for coherent detection at the receiver.

The coherent demodulator is described in the next section. The phase information generated by the coherent modulator should be passed to the coherent demodulator function (bfsk_coherent_demod) shown in 2.12.5. If noncoherent BFSK is chosen at the modulator, the phase argument returned by the function is irrelevant at the receiver. Moreover, a coherently modulated BFSK signal can also be noncoherently demodulated. The noncoherent demodulator is given in section 2.12.6.

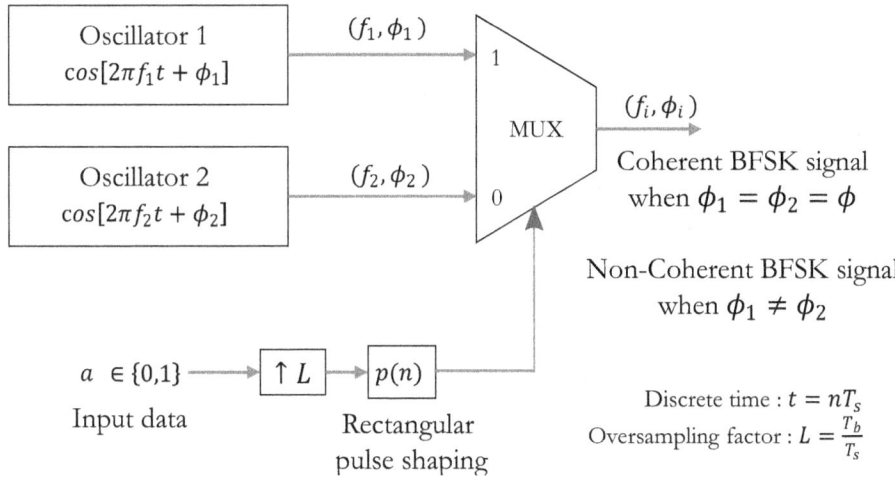

Fig. 2.38: Discrete-time equivalent model for BFSK modulation

Program 2.30: *bfsk_mod.m*: Function for generating coherent and non-coherent discrete-time BFSK signal

```matlab
function [s,t,phase,at] = bfsk_mod(a,Fc,Fd,L,Fs,fsk_type)
%Function to modulate an incoming binary stream using BFSK
%a - input binary data stream (0's and 1's) to modulate
%Fc - center frequency of the carrier in Hertz
%Fd - frequency separation measured from Fc
%L - number of samples in 1-bit period
%Fs - Sampling frequency for discrete-time simulation
%fsk_type - 'COHERENT' (default) or 'NONCOHERENT' FSK generation
%s - BFSK modulated signal
%t - generated time base for the modulated signal
%phase - initial phase generated by modulator, applicable only for coherent FSK.
%It can be used when using coherent detection at Rx
%at - data waveform for the input data
phase=0;
at = kron(a,ones(1,L)); %data to waveform
t = (0:1:length(at)-1)/Fs; %time base
if strcmpi(fsk_type,'NONCOHERENT'),
  c1 = cos(2*pi*(Fc+Fd/2)*t+2*pi*rand);%carrier 1 with random phase
  c2 = cos(2*pi*(Fc-Fd/2)*t+2*pi*rand);%carrier 2 with random phase
else
  phase=2*pi*rand;%random phase from uniform distribution [0,2pi)
  c1 = cos(2*pi*(Fc+Fd/2)*t+phase);%carrier 1 with random phase
  c2 = cos(2*pi*(Fc-Fd/2)*t+phase);%carrier 2 with random phase
end
s = at.*c1 +(-at+1).*c2; %BFSK signal (MUX selection)
doPlot=0;
if doPlot,
  figure;subplot(2,1,1);plot(t,at);subplot(2,1,2);plot(t,s);
end;
```

2.12.5 Coherent Demodulator

There are several ways to demodulate a coherent FSK signal and the *Voltage Controlled Oscillator* (VCO) based demodulator is the most popular among them. The coherent demodulator for FSK can also be implemented using only one correlator, and its discrete-time equivalent model is shown in Figure 2.39. Here, the demodulator requires the initial phase information ($\phi_1 = \phi_2 = \phi$) that was used to generate the coherent FSK signal at the transmitter. If this phase information is not accurately known at the receiver, the demodulator fails completely. In order to avoid carrier phase recovery, a non-coherent demodulator can be used to demodulate a coherent FSK signal, but there is an error rate performance penalty in doing so (see next section).

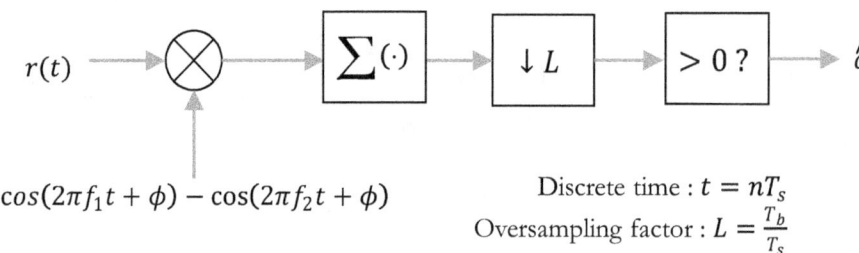

$$cos(2\pi f_1 t + \phi) - cos(2\pi f_2 t + \phi)$$

Discrete time : $t = nT_s$

Oversampling factor : $L = \dfrac{T_b}{T_s}$

Fig. 2.39: Discrete-time equivalent model for coherent demodulation of coherent FSK

The theoretical bit-error probability of a coherently detected coherent binary FSK, over AWGN noise, where two carrier signals are orthogonal and equi-probable, is given by

$$P_b = Q\left(\sqrt{\frac{E_b}{N_0}}\right) = \frac{1}{2}\, erfc\left(\sqrt{\frac{E_b}{2N_0}}\right) \tag{2.59}$$

Program 2.31: *bfsk_coherent_demod.m*: Coherent demodulator for demodulating a coherent BFSK signal

```
1  function a_cap = bfsk_coherent_demod(r,Fc,Fd,L,Fs,phase)
2  %Coherent demodulation of BFSK modulated signal
3  %r - BFSK modulated signal at the receiver
4  %Fc - center frequency of the carrier in Hertz
5  %Fd - frequency separation measured from Fc
6  %L - number of samples in 1-bit period
7  %Fs - Sampling frequency for discrete-time simulation
8  %phase - initial phase generated at the transmitter
9  %a_cap - data bits after demodulation
10 t = (0:1:length(r)-1)/Fs; %time base
11 x = r.*(cos(2*pi*(Fc+Fd/2)*t+phase)-cos(2*pi*(Fc-Fd/2)*t+phase));
12 y = conv(x,ones(1,L)); %integrate from 0 to Tb
13 a_cap = y(L:L:end)>0;%sample at every sampling instant and detect
14 end
```

2.12.6 Non-coherent Demodulator

A non-coherent FSK demodulator does not care about the initial phase information used at transmitter, hence avoids the need for carrier phase recovery at the receiver. As a result, the non-coherent demodulator is easier to build and therefore offers a better solution for practical applications.

The discrete-time equivalent model for a correlator/square-law based non-coherent demodulator is shown in Figure 2.40. It can be used to demodulate both coherent FSK and non-coherent FSK signal.

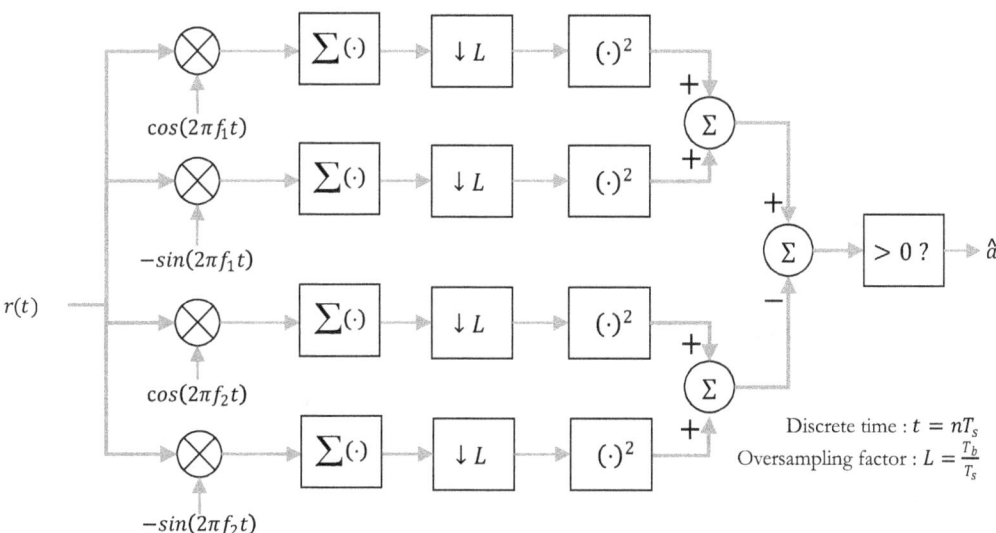

Fig. 2.40: Discrete-time equivalent model for square-law based noncoherent demodulation for coherent/non-coherent FSK signal

The theoretical bit-error probability of non-coherently detected binary FSK over AWGN noise, where two carrier signals are orthogonal and equi-probable, is given by

$$P_b = \frac{1}{2} e^{E_b/(2N_0)} \tag{2.60}$$

Program 2.32: *bfsk_noncoherent_demod.m*: Square-law based non-coherent demodulator

```
1  function a_cap = bfsk_noncoherent_demod(r,Fc,Fd,L,Fs)
2  %Non-coherent demodulation of BFSK modulated signal
3  %r - BFSK modulated signal at the receiver
4  %Fc - center frequency of the carrier in Hertz
5  %Fd - frequency separation measured from Fc
6  %L - number of samples in 1-bit period
7  %Fs - Sampling frequency for discrete-time simulation
8  %a_cap - data bits after demodulation
9  t = (0:1:length(r)-1)/Fs; %time base
10 F1 = (Fc+Fd/2); F2 = (Fc-Fd/2);
11
12 %define four basis functions
13 p1c = cos(2*pi*F1*t); p2c = cos(2*pi*F2*t);
```

```
14    p1s = -sin(2*pi*F1*t); p2s = -sin(2*pi*F2*t);
15
16    %multiply and integrate from 0 to Tb
17    r1c = conv(r.*p1c,ones(1,L)); r2c = conv(r.*p2c,ones(1,L));
18    r1s = conv(r.*p1s,ones(1,L)); r2s = conv(r.*p2s,ones(1,L));
19
20    %sample at every sampling instant
21    r1c = r1c(L:L:end); r2c = r2c(L:L:end);
22    r1s = r1s(L:L:end); r2s = r2s(L:L:end);
23
24    %square and add
25    x = r1c.^2 + r1s.^2;
26    y = r2c.^2 + r2s.^2;
27
28    a_cap=(x-y)>0; %compare and decide
```

2.12.7 Performance simulation

The error rate performances of coherent and non coherent demodulation of BFSK signal over an AWGN channel is simulated and the results are plotted in Figure 2.41. The simulation code performs BFSK modulation and demodulation by utilizing various functions described in the previous sections: bfsk_mod for generating BFSK modulated signal, bfsk_coherent_demod for coherent demodulation and for non-coherent demodulation the function bfsk_noncoherent_demod is used.

In the code sample below, the FSK type at the transmitter is chosen as coherent. This generates a coherent BFSK signal at the transmitter. The generated signal is then passed through an AWGN channel. The received signal is then independently demodulated using a coherent demodulator and a non-coherent demodulator. For both the cases, the error rate performance is evaluated over a range of E_b/N_0 values.

On the other hand, if the FSK type is chosen as non-coherent in the modulator, the coherent demodulation is not applicable. Therefore, for this case only the performance of the non-coherent demodulator will be plotted.

Program 2.33: *bfsk_wfm_sim.m*: Simulating performance of coherent and non-coherent BFSK demodulation

```
1     %Eb/N0 Vs BER for BFSK mod/coherent-demod (waveform simulation)
2     clearvars; clc;
3     N=100000;%Number of bits to transmit
4     EbN0dB = -4:2:10; % Eb/N0 range in dB for simulation
5     Fc = 400; %center carrier frequency f_c- integral multiple of 1/Tb
6     fsk_type = 'COHERENT'; %COHERENT/NONCOHERENT FSK generation at Tx
7     h = 1; %modulation index
8     % h should be minimum 0.5 for coherent FSK or multiples of 0.5
9     %h should be minimum 1 for non-coherent FSK or multiples of 1
10    L = 40; %oversampling factor
11    Fs = 8*Fc; %sampling frequency for discrete-time simulation
12    Tb = L/Fs;% bit period, f_c is integral multiple of 1/Tb
13    Fd = h/Tb;%Frequency separation
14
15    EbN0lin = 10.^(EbN0dB/10); %converting dB values to linear scale
16    BER_coherent = zeros(length(EbN0dB),1); %BER values for each Eb/N0
```

```
17   BER_nonCoherent = BER_coherent;
18
19   %-----------------Transmitter---------------------
20   a = rand(1,N)>0.5; %random bits - input to BFSK modulator
21   [s,t,phase]=bfsk_mod(a,Fc,Fd,L,Fs,fsk_type); %BFSK modulation
22
23   for i=1:length(EbN0dB),
24     Eb=L*sum(abs(s).^2)/(length(s)); %compute energy per bit
25     N0= Eb/EbN0lin(i); %required noise spectral density from Eb/N0
26     n = sqrt(N0/2)*(randn(1,length(s)));%computed noise
27     r = s + n;%add noise
28
29     %--------------Receiver-------------------
30     if strcmpi(fsk_type,'COHERENT'),
31       %coherent FSK could be demodulated coherently or non-coherently
32       a_cap1 = bfsk_coherent_demod(r,Fc,Fd,L,Fs,phase);%coherent demod
33       a_cap2 = bfsk_noncoherent_demod(r,Fc,Fd,L,Fs);%noncoherent demod
34       BER_coherent(i) = sum(a~=a_cap1)/N;%BER for coherent case
35       BER_nonCoherent(i) = sum(a~=a_cap2)/N;%BER for non-coherent case
36     end
37
38     if strcmpi(fsk_type,'NONCOHERENT'),
39       %non-coherent FSK can only non-coherently demodulated
40       a_cap2 = bfsk_noncoherent_demod(r,Fc,Fd,L,Fs);%noncoherent demod
41       BER_nonCoherent(i) = sum(a~=a_cap2)/N;%BER for non-coherent case
42     end
43   end
44
45   %------Theoretical Bit Error Rate-------------
46   coherent = 0.5*erfc(sqrt(EbN0lin/2));%Theory BER - coherent case
47   nonCoherent = 0.5*exp(-EbN0lin/2);%Theory BER - non-coherent case
48   figure; %-------------Plot performance curve----------------------
49   if strcmpi(fsk_type,'COHERENT'),
50     semilogy(EbN0dB,BER_coherent,'k*'); hold on; %sim BER
51     semilogy(EbN0dB,BER_nonCoherent,'m*'); %sim BER
52     semilogy(EbN0dB,coherent,'r-');
53     semilogy(EbN0dB,nonCoherent,'b-');
54     title('Probability of Bit Error for coherent BFSK modulation');
55     legend('Sim-coherent demod','Sim-noncoherent demod','theory-coherent demod','
          theory-noncoherent demod');
56   end
57
58   if strcmpi(fsk_type,'NONCOHERENT'),
59     semilogy(EbN0dB,BER_nonCoherent,'m*'); hold on;%sim BER
60     semilogy(EbN0dB,nonCoherent,'b-');
61     title('Probability of Bit Error for non-coherent BFSK modulation');
62     legend('Sim-noncoherent demod','theory-noncoherent demod');
63   end
64   xlabel('E_b/N_0 (dB)');ylabel('Probability of Bit Error - P_b');
```

Fig. 2.41: Performance of coherent BFSK when demodulated using coherent and non-coherent techniques

2.12.8 Power spectral density

Assuming rectangular pulse shaping at the baseband, the power spectral density plot for the BFSK modulated signal is simulated using the windowed Welch spectrum estimation method described in chapter chapter 1 section 1.4 and the result is plotted in Figure 2.42. For this simulation, the center frequency of the carrier is chosen as $f_c = 400Hz$, sampling frequency $f_s = 8 \times f_c$, modulation index $h = 1$ and bit period $T_b = 0.0125$. As expected for this configuration, the PSD plot reveals two spikes at frequencies $f_1 = 440Hz$ and $f_2 = 360Hz$ that are separated from each other by $\Delta f = 80Hz$.

Fig. 2.42: BFSK power spectra

References

1. Lloyd N. Trefethen, David Bau III (1997), *Numerical linear algebra*, Philadelphia: Society for Industrial and Applied Mathematics, ISBN 978-0-89871-361-9, pp.56
2. Park, J. H., Jr., *On binary DPSK detection*,IEEE Trans. Communication, vol. 26, no. 4, April 1978,pp. 484–486
3. Fuqin Xiong,*Digital Modulation Techniques*, Second Edition, Artech House, Inc. ISBN-1580538630
4. S. A. Rhodes,*Effects of hardlimiting on bandlimited transmissions with conventional and offset QPSK modulation*, in Proc. Nat. Telecommun. Conf., Houston, TX, 1972, PP. 20F/1-20F/7
5. S. A. Rhodes,*Effect of noisy phase reference on coherent detection of offset QPSK signals*, IEEE Trans. Commun., vol. COM-22, PP. 1046-1055, Aug. 1974
6. S. Pasupathy, *Minimum Shift Keying: A Spectrally Efficient Modulation*, IEEE Communications Magazine, vol. 17, no. 4, pp. 14-22, July 1979
7. Dayan Adionel Guimaraes, *Contributions to the Understanding of the MSK Modulation*, revista telecomunicaes, VOL. 11, NO. 01, MAIO DE 2008.
8. Clay S. Turner, *Raised Cosine and Root Raised Cosine Formulae*, Wireless Systems Engineering, Inc, (May 29, 2007) V1.2
9. Murota, K. and Hirade, K., *GMSK Modulation for Digital Mobile Radio Telephony*, IEEE Transactions on Communications, vol COM-29, No. 7. pp. 1044-1050, July 1981.
10. Marvin K. Simon, *Bandwidth-Efficient Digital Modulation with Application to Deep Space Communications*, JPL Deep Space Communications and Navigation Series,Wiley-Interscience, Hoboken, New Jersey, 2003, ISBN 0-471-44536-3,pp-57.
11. Bridges Paul A, *Digital demodulation techniques for radio channels*, Master thesis, Victoria University of Technology, 1993. http://vuir.vu.edu.au/15572/
12. J. B. Anderson, T. Aulin, and C.-E. Sundberg, *Digital Phase Modulation*, ISBN 978-0306421952, Plenum Press. New York. 1986

Chapter 3
Digital Modulators and Demodulators - Complex Baseband Equivalent Models

Abstract In chapter 2, we saw how waveform level simulations were used to simulate the performance of various modulation techniques. In such simulation models, every cycle of the sinusoidal carrier gets simulated, eventually consuming more memory and time. On the other hand, the complex baseband equivalent models, the subject of this chapter, operates on sample basis. This approach is much simpler to implement. It offers a great advantage by drastically reducing memory requirements as well as yielding results in a much shorter timespan. The focus of this chapter is on developing simulation codes for implementing various modulations using the complex baseband equivalent approach.

3.1 Introduction

The *passband model* and *equivalent baseband model* are fundamental models for simulating a communication system. In the passband model, also called as *waveform simulation model*, the transmitted signal, channel noise and the received signal are all represented by samples of waveforms. Since every detail of the RF carrier gets simulated, it consumes more memory and time.

In the case of discrete-time equivalent baseband model, only the value of a symbol at the symbol-sampling time instant is considered. Therefore, it consumes less memory and yields results in a very short span of time when compared to the passband models. Such models operate near zero frequency, suppressing the RF carrier and hence the number of samples required for simulation is greatly reduced. They are more suitable for performance analysis simulations. If the behavior of the system is well understood, the model can be simplified further.

3.2 Complex baseband representation of modulated signal

By definition, a passband signal is a signal whose one-sided energy spectrum is centered on non-zero carrier frequency f_c and does not extend to DC. A passband signal or any digitally modulated RF waveform is represented as

$$\tilde{s}(t) = a(t)cos\left[2\pi f_c t + \phi(t)\right] = s_I(t)cos(2\pi f_c t) - s_Q(t)sin(2\pi f_c t) \tag{3.1}$$

where,

$$a(t) = \sqrt{s_I(t)^2 + s_Q(t)^2} \text{ and } \phi(t) = tan^{-1}\left(\frac{s_Q(t)}{s_I(t)}\right) \tag{3.2}$$

Recognizing that the sine and cosine terms in the equation 3.1 are orthogonal components with respect to each other, the signal can be represented in complex form as

$$s(t) = s_I(t) + js_Q(t) \qquad (3.3)$$

When represented in this form, the signal $s(t)$ is called the *complex envelope* or the *complex baseband equivalent representation* of the real signal $\tilde{s}(t)$. The components $s_I(t)$ and $s_Q(t)$ are called *inphase* and *quadrature* components respectively. Comparing equations 3.1 and 3.3, it is evident that in the complex baseband equivalent representation, the carrier frequency is suppressed. This greatly reduces both the sampling frequency requirements and the memory needed for simulating the model. Furthermore, equation 3.1, provides a practical way to convert a passband signal to its baseband equivalent and vice-versa [1], as illustrated in Figure 3.1[†]

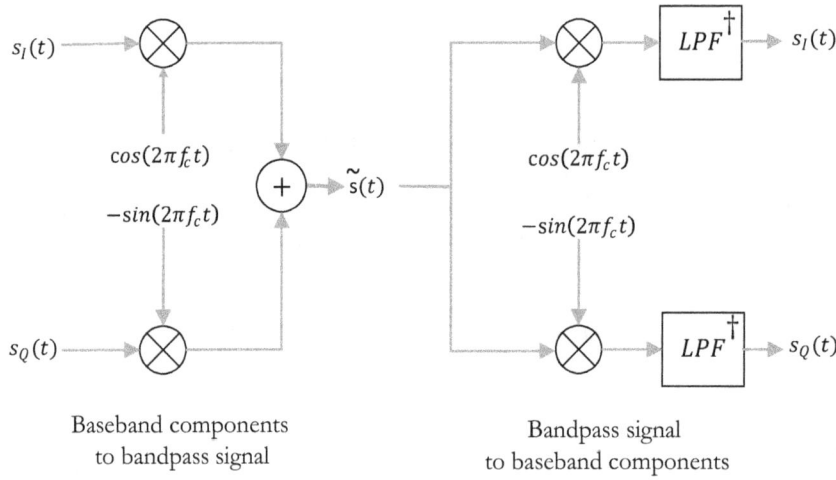

Baseband components   Bandpass signal
to bandpass signal   to baseband components

Fig. 3.1: Conversion from baseband to passband and vice-versa

3.3 Complex baseband representation of channel response

In the typical communication system model shown in Figure 3.2(a), the signals represented are real passband signals. The digitally modulated signal $\tilde{s}(t)$ occupies a band-limited spectrum around the carrier frequency (f_c). The channel $\tilde{h}(t)$ is modeled as a *linear time invariant* system which is also band-limited in nature. The effect of the channel on the modulated signal is represented as *linear convolution* (denoted by the $*$ operator). Then, the received signal is given by

$$\tilde{y} = \tilde{s} * \tilde{h} + \tilde{n} \qquad (3.4)$$

The corresponding complex baseband equivalent (Figure 3.2(b)) is expressed as

$$(y_I + jy_Q) = (s_I + js_Q) * (h_I + jh_Q) + (n_I + jn_Q) \qquad (3.5)$$

[†] LPF = Low Pass Filter

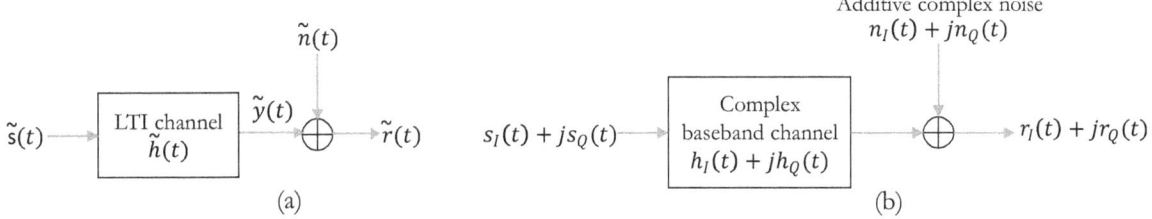

Fig. 3.2: Passband channel model and its baseband equivalent

3.4 Modulators for amplitude and phase modulations

In the following sections, the discrete-time baseband models for the following digital modulators are implemented as modular functions.

- Pulse Amplitude Modulation (PAM) - mpam_modulation.m
- Pulse Shift Keying Modulation (PSK) - mpsk_modulation.m
- Quadrature Amplitude Modulation (QAM) - mqam_modulation.m
- Orthogonal Frequency Shift Keying Modulation (FSK) - mfsk_modulation.m

To maintain continuity, only the first three modulators are described in the section that follows. The modulator for MFSK modulation is described separately in section 3.6.

The drafted functions will have a common format and we can draft a wrapper function to incorporate all the models as shown below. The wrapper function becomes handy and can be incorporated in bigger scheme of implementations.

Program 3.1: *modulate.m*: Function to modulate information using a selected modulation technique

```
function [s,ref]=modulate(MOD_TYPE,M,d,COHERENCE)
%Wrapper function to call various digital modulation techniques
%  MOD_TYPE - 'PSK','QAM','PAM','FSK'
%  M - modulation order, For BPSK M=2, QPSK M=4, 256-QAM M=256 etc..,
%  d - data symbols to be modulated drawn from the set {1,2,...,M}
%  COHERENCE - only applicable if FSK modulation is chosen
%            - 'coherent' for coherent MFSK
%            - 'noncoherent' for coherent MFSK
%  s - modulated symbols
%  ref - ideal constellation points that could be used by an IQ detector
switch lower(MOD_TYPE)
    case 'bpsk'
        [s,ref] = mpsk_modulator(2,d);
    case 'psk'
        [s,ref] = mpsk_modulator(M,d);
    case 'qam'
        [s,ref] = mqam_modulator(M,d);
    case 'pam'
        [s,ref] = mpam_modulator(M,d);
    case 'fsk'
        [s,ref] = mfsk_modulator(M,d,COHERENCE);
    otherwise
```

```
23       error('Invalid Modulation specified');
24   end
```

3.4.1 Pulse Amplitude Modulation (M-PAM)

MPAM modulation is a one dimensional modulation technique that has no quadrature component ($s_Q = 0$). All the information gets encoded in the signal amplitude. A discrete-time baseband model for an M-PAM modulator, transmits a series of information symbols drawn from the set $m \in \{1, 2, ..., M\}$ with each transmitted symbol holding k bits of information ($k = log_2(M)$). The information symbols are digitally modulated using M-PAM signaling. The general expression for generating the M-PAM signal constellation set is given by

$$A_m = 2m - 1 - M, \quad m = 1, 2, ..., M \tag{3.6}$$

Here, M denotes the *modulation order* and it defines the number of points in the *ideal reference* constellation. The value of M depends on a parameter k – the number of bits we wish to squeeze in a single M-PAM symbol. For example, if 3 bits ($k = 3$) are squeezed in one transmit symbol, $M = 2^K = 2^3 = 8$, that results in 8-PAM configuration. For, the M-PAM signaling, the constellation points are located at $\pm 1, \pm 3, ..., \pm (M - 1)$, as shown in Figure 3.3. The discrete-time baseband equivalent model for an MPAM modulator is coded as a function.

Program 3.2: *mpam_modulator.m*: MPAM modulator

```
1   function [s,ref]=mpam_modulator(M,d)
2   %Function to MPAM modulate the vector of data symbols - d
3   %[s,ref]=mpam_modulator(M,d) modulates the symbols defined by the
4   %vector d using MPAM modulation, where M specifies the order of
5   %M-PAM modulation and the vector d contains symbols whose values
6   %in the range 1:M. The output s is the modulated output and ref
7   %represents the reference constellation
8   m=1:1:M;
9   Am=complex(2*m-1-M); %All possibe amplitude levels
10  s = complex(Am(d)); %M-PAM transmission
11  ref = Am; %reference constellation
12  end
```

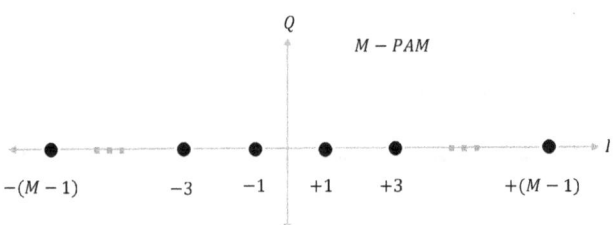

Fig. 3.3: M-PAM constellation

3.4.2 Phase Shift Keying Modulation (M-PSK)

In phase shift keying, all the information gets encoded in the phase of the carrier signal. The M-PSK modulator transmits a series of information symbols drawn from the set $m \in \{1, 2, \ldots, M\}$. Each transmitted symbol holds k bits of information ($k = log_2(M)$). The information symbols are digital modulated using M-PSK mapping. The general expression for generating the M-PSK signal set is given by

$$s_m(t) = A cos \left[2\pi f_c t - \frac{(m-1)2\pi}{M} \right], \quad m = 1, 2, \ldots, M \tag{3.7}$$

Here, M denotes the modulation order and it defines the number of constellation points in the reference constellation. The value of M depends on the parameter k – the number of bits we wish to squeeze in a single M-PSK symbol. For example if we wish to squeeze in 3 bits ($k = 3$) in one transmit symbol, then $M = 2^k = 2^3 = 8$ and this results in 8-PSK configuration. $M = 2$ gives BPSK (Binary Phase Shift Keying) configuration. The configuration with $M = 4$ is referred as QPSK (Quadrature Phase Shift Keying). The parameter A is the amplitude scaling factor. Using trigonometric identity, equation 3.7 can be separated into cosine and sine basis functions as follows

$$s_m(t) = A cos \left[\frac{(m-1)2\pi}{M} \right] cos(2\pi f_c t) + A sin \left[\frac{(m-1)2\pi}{M} \right] sin(2\pi f_c t), \quad m = 1, 2, \ldots, M \tag{3.8}$$

This can be expressed as a combination of in-phase and quadrature phase components on an I-Q plane as

$$s_m(t) = A cos \left[\frac{(m-1)2\pi}{M} \right] + jA sin \left[\frac{(m-1)2\pi}{M} \right], \quad m = 1, 2, \ldots, M \tag{3.9}$$

Normalizing the amplitude as $A = 1/\sqrt{2}$, the points on the reference constellation will be placed on the unit circle. The MPSK modulator is constructed based on this equation and the ideal constellations for $M = 4, 8$ and 16 PSK modulations are shown in Figure 3.4.

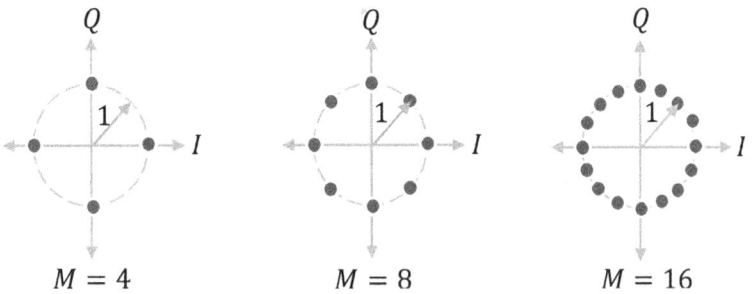

$M = 4 \qquad\qquad M = 8 \qquad\qquad M = 16$

Fig. 3.4: M-PSK constellations

Program 3.3: *mpsk_modulator.m*: MPSK modulator

```
1  function [s,ref]=mpsk_modulator(M,d)
2  %Function to MPSK modulate the vector of data symbols - d
3  %[s,ref]=mpsk_modulator(M,d) modulates the symbols defined by the
4  %vector d using MPSK modulation, where M specifies the order of
5  %M-PSK modulation and the vector d contains symbols whose values
6  %in the range 1:M. The output s is the modulated output and ref
```

```
7   %represents the reference constellation that can be used in demod
8   ref_i= 1/sqrt(2)*cos(((1:1:M)-1)/M*2*pi);
9   ref_q= 1/sqrt(2)*sin(((1:1:M)-1)/M*2*pi);
10  ref = ref_i+1i*ref_q;
11  s = ref(d); %M-PSK Mapping
12  end
```

3.4.3 Quadrature Amplitude Modulation (M-QAM)

In MQAM modulations, the information bits are encoded as variations in the amplitude and the phase of the signal. The M-QAM modulator transmits a series of information symbols drawn from the set $m \in \{1, 2, .., M\}$, with each transmitted symbol holding k bits of information ($k = log2(M)$). To restrict the erroneous receiver decisions to single bit errors, the information symbols are Gray coded. The information symbols are then digitally modulated using a rectangular M-QAM technique, whose signal set is given by

$$s = a + jb \quad where \ a, b \in \{\pm 1, \pm 3, \ldots, \pm (\lceil \sqrt{M} \rceil - 1)\} \tag{3.10}$$

Karnaugh-map walks and Gray codes

In any M-QAM constellation, in order to restrict the erroneous symbol decisions to single bit errors, the adjacent symbols in the transmitter constellation should not differ by more than one bit. This is usually achieved by converting the input symbols to Gray coded symbols and then mapping it to the desired QAM constellation. But this intermediate step can be skipped altogether by using a Look-Up-Table (LUT) approach which properly translates the input symbol to appropriate position in the constellation.

We will exploit the inherent property of Karnaugh Maps to generate the look-up table of dimension $N \times N$ (where $N = \sqrt{M}$) for the Gray coded MQAM constellation which is rectangular and symmetric ($M = 4, 16, 64, 256, \cdots$). The first step in constructing a QAM constellation is to convert the sequential numbers representing the message symbols to Gray coded format. The function to convert decimal numbers to Gray codes is given next.

Program 3.4: *dec2gray.m*: Decimal to Gray code converter

```
1   function [grayCoded]=dec2gray(decInput)
2   %convert decimal to Gray code representation
3   %example: x= [0 1 2 3 4 5 6 7] %decimal
4   %y=dec2gray(x); %returns y = [ 0 1 3 2 6 7 5 4] %Gray coded
5   [rows,cols]=size(decInput);
6   grayCoded=zeros(rows,cols);
7   for i=1:rows
8     for j=1:cols
9       grayCoded(i,j)=bitxor(bitshift(decInput(i,j),-1),decInput(i,j));
10    end
11  end
```

If you are familiar with Karnaugh maps (K-Maps), it is easier for you to identify that the K-Maps are constructed based on Gray Codes. By the nature of the construction of K-Maps, the address of the adjacent cells differ by only one bit. If we superimpose the given M-QAM constellation on the K-Map and walk through the address of each cell in a certain pattern, it gives the Gray-coded M-QAM constellation.

As mentioned, a walk through the K-Map will produce a sequence of Gray codes. Moreover, if the walk can be looped back to the origin or starting point, it will generate a sequence of cyclic Gray codes. Different walking patterns are possible on K-Maps that generate different sequences of Gray codes. Some of the walks on a 4 × 4 K-Map are shown in the Figure 3.5. This can be readily extended to any K-Map configuration of higher order.

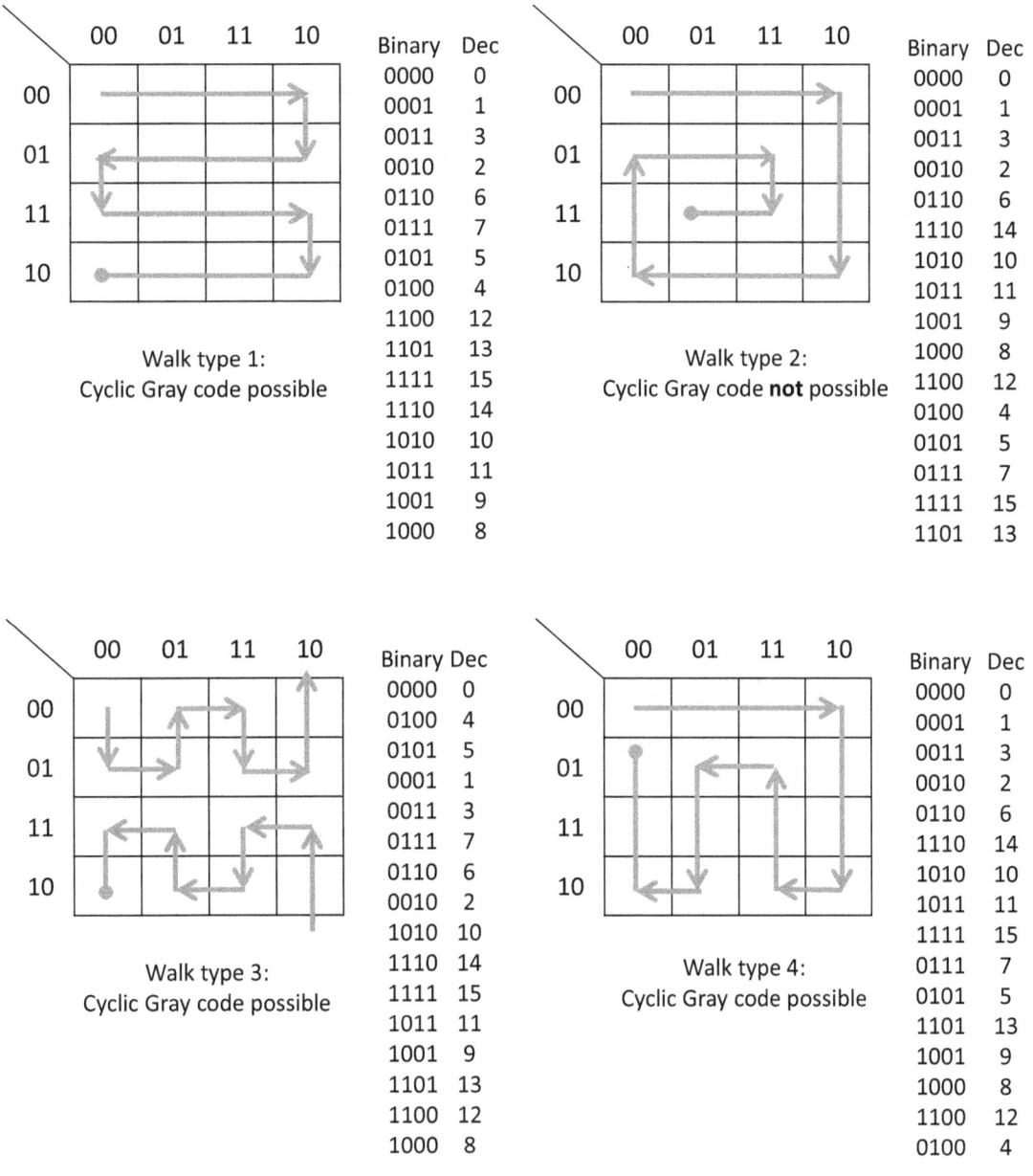

Fig. 3.5: Karnaugh Map walks and Gray codes

In walk types 1,3 and 4, the address of the starting point and end point differ by only one bit and the corresponding cells are adjacent to each other. In effect, the walk can be looped to give cyclic Gray codes.

But in type 2 walk, the starting cell (0000) and the ending cell (1101) are not adjacent to each other and thus the Gray code generated using this pattern of walk is not cyclic. Thus far, type 1 walk is the simplest. All we have to do is alternate the direction of the walk for every row and read the Gray coded address. The Matlab function `constructQAM.m` given in the next subsection implements the walk type 1 for constructing a MQAM constellation.

Rectangular QAM from PAM constellation

There exist other constellation shapes (like circular, triangular constellations) that are more efficient than the standard rectangular constellation [2]. Symmetric rectangular (square) constellations are the preferred choice of implementation due to their simplicity in implementing modulation and demodulation functions.

Any rectangular QAM constellation is equivalent to superimposing two PAM signals on quadrature carriers. For example, 16-QAM constellation points can be generated from two 4-PAM signals, similarly the 64-QAM constellation points can be generated from two 8-PAM signals. The generic equation to generate PAM signals of dimension D is

$$A_d = 2d - 1 - D \quad ; d = 1, 2, \ldots, D \tag{3.11}$$

For generating 16-QAM, the dimension D of PAM is set to $D = \sqrt{16} = 4$. Thus for constructing a M-QAM constellation, the PAM dimension is set as $D = \sqrt{M}$. Matlab code for dynamically generating M-QAM constellation points based on Karnaugh map Gray code walk is given below. The resulting ideal constellations for Gray coded 16-QAM and 64-QAM are shown in Figure 3.6.

Program 3.5: *constructQAM.m*: Constructing Gray coded MQAM constellation

```
1   function [ref,varargout]= constructQAM(M)
2   %Function to construct gray codes symbol constellation for M-QAM
3   % [ref]=constructQAM(M) - returns the ideal signaling points (ref)
4   % in a symmetric rectangular M-QAM constellation, where M is the
5   % level of QAM modulation. The returned constellation points are
6   % arranged such that the index of the points are arranged in a
7   % Gray-coded manner. When plotted, indices of constellation points
8   % will differ by 1 bit.
9   %
10  % [ref,I,Q]=constructQAM(M) - returns the ideal signaling points
11  % (ref) along with the IQ components breakup in a symmetric
12  % rectangular M-QAM constellation, where M is the level of QAM
13  % modulation. The returned constellation points are arranged such
14  % that the index of the points are arranged in a Gray-coded manner.
15  n=0:1:M-1; %Sequential address from 0 to M-1 (1xM dimension)
16
17  %------Addresses in Kmap - Gray code walk---------------
18  a=dec2gray(n); %Convert linear addresses to gray code
19  N=sqrt(M); %Dimension of K-Map - N x N matrix
20  a=reshape(a,N,N).'; %NxN gray coded matrix
21  evenRows=2:2:size(a,1); %identify alternate rows
22  a(evenRows,:)=fliplr(a(evenRows,:));%Flip rows - KMap representation
23  nGray=reshape(a.',1,M); %reshape to 1xM - Gray code walk on KMap
24
25  %Construction of ideal M-QAM constellation from sqrt(M)-PAM
26  D=sqrt(M); %Dimension of PAM constellation
27  x=floor(nGray/D);
```

```
28   y=mod(nGray,D);
29   Ax=2*(x+1)-1-D; %PAM Amplitudes 2m-1-D - real axis
30   Ay=2*(y+1)-1-D; %PAM Amplitudes 2m-1-D - imag axis
31   ref=Ax+1i*Ay; %assign complex numbers to reference
32   if nargout==2 %if only one variable argument is given
33       varargout{1}=Ax; %Real part (I)
34   elseif nargout==3 %if two variable arguments are given
35       varargout{1}=Ax; %Real part (I)
36       varargout{2}=Ay; %Imaginary part (Q)
37   end
```

The wrapper function for implementing an MQAM modulator is given next. The function returns both the modulated symbols (s) and the reference constellation points that could be used in the receiver for demodulation. The signal space constellation that is generated using the following function is given in Figure 3.6.

Program 3.6: *mqam_modulator.m*: MQAM modulator

```
1    function [s,ref]=mqam_modulator(M,d)
2    %Function to MQAM modulate the vector of data symbols - d
3    %[s,ref]=mqam_modulator(M,d) modulates the symbols defined by the
4    %vector d using MQAM modulation, where M specifies the order of
5    %M-QAM modulation and the vector d contains symbols whose values
6    %range 1:M. The output s is the modulated output and ref
7    %represents the reference constellation that can be used in demod
8    if(((M~=1) && ~mod(floor(log2(M)),2))==0), %M not a even power of 2
9        error('Only Square MQAM supported. M must be even power of 2');
10   end
11     ref=constructQAM(M); %construct reference constellation
12     s=ref(d); %map information symbols to modulated symbols
13   end
```

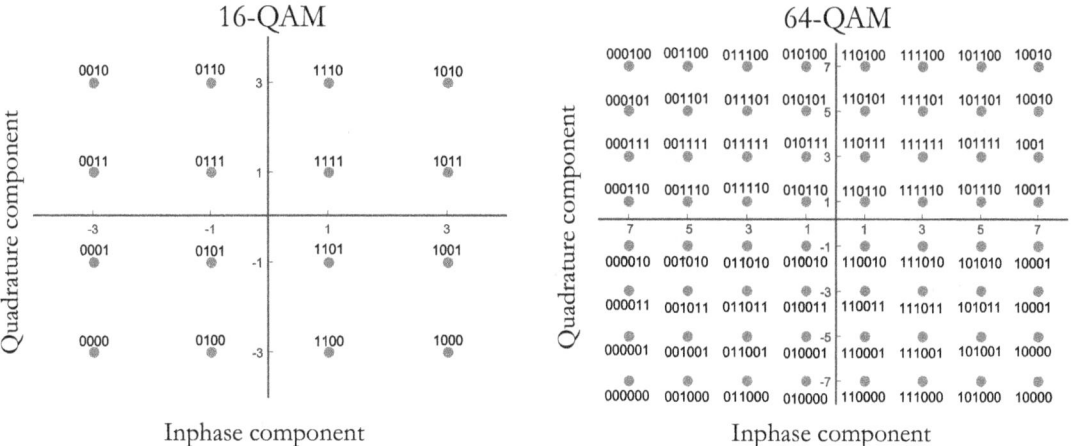

Fig. 3.6: Signal space constellations for 16-QAM and 64-QAM

3.5 Demodulators for amplitude and phase modulations

The discrete-time baseband models for various digital demodulators are introduced and drafted as functions for implementation in Matlab. We will implement demodulators (with function names) for the following digital modulation schemes. To maintain continuity, only the first three detection techniques are described in the subsection that follows. The detector for MFSK demodulation is described separately in 3.6.

- PAM detection - mpam_detector.m
- PSK detection - mpsk_detector.m
- QAM detection - mqam_detector.m
- FSK detection - mfsk_detector.m

The drafted functions will have a common format and we can draft a wrapper function to incorporate all the models as shown below. The wrapper function comes handy when incorporated in bigger scheme of implementations.

Program 3.7: *demodulate.m*: Function to demodulate using a selected demodulation technique

```
function [dCap]=demodulate(MOD_TYPE,M,r,COHERENCE)
%Wrapper functin to call various digital demodulation techniques
% MOD_TYPE - 'PSK','QAM','PAM','FSK'
% M - modulation order, For BPSK M=2, QPSK M=4, 256-QAM M=256 etc..,
% COHERENCE - only applicable if FSK modulation is chosen
%            - 'coherent' for coherent MFSK
%            - 'noncoherent' for coherent MFSK
% r - received modulated symbols
% dCap - demodulated information symbols

%Construct the reference constellation for the selected mod. type
switch lower(MOD_TYPE)
    case 'bpsk'
        M=2;
        dCap= mpsk_detector(2,r);
    case 'psk'
        dCap= mpsk_detector(M,r);
    case 'qam'
        dCap= mqam_detector(M,r);
    case 'pam'
        dCap= mpam_detector(M,r);
    case 'fsk'
        dCap= mfsk_detector(M,r,COHERENCE);
    otherwise
        error('Invalid Modulation specified');
end
```

3.5.1 M-PAM detection

The IQ detection technique based on minimum Euclidean distance metric (described in section 3.5.4) can be leveraged for implementing a coherent detector for **MPAM** detection.

Program 3.8: *mpam_detector.m*: Coherent detection of MPAM signaling points

```matlab
function [dCap]= mpam_detector(M,r)
%Function to detect MPAM modulated symbols
%[dCap]= mqam_detector(M,r) detects the received MPAM signal points
%points - 'r'. M is the modulation level of MPAM
m=1:1:M; Am=complex(2*m-1-M); %all possibe amplitude levels
ref = Am; %reference constellation for MPAM
[~,dCap]= iqOptDetector(r,ref); %IQ detection
end
```

3.5.2 M-PSK detection

The IQ detection technique based on minimum Euclidean distance metric (described in section 3.5.4) can be leveraged for implementing a coherent detector for MPSK detection.

Program 3.9: *mpsk_detector.m*: Coherent detection of MPSK signaling points

```matlab
function [dCap]= mpsk_detector(M,r)
%Function to detect MPSK modulated symbols
%[dCap]= mpsk_detector(M,r) detects the received MPSK signal points
%points - 'r'. M is the modulation level of MPSK
ref_i= 1/sqrt(2)*cos(((1:1:M)-1)/M*2*pi);
ref_q= 1/sqrt(2)*sin(((1:1:M)-1)/M*2*pi);
ref = ref_i+1i*ref_q; %reference constellation for MPSK
[~,dCap]= iqOptDetector(r,ref); %IQ detection
end
```

3.5.3 M-QAM detection

The IQ detection technique based on minimum Euclidean distance metric (described in section 3.5.4) can be leveraged for implementing a coherent detector for MQAM detection.

Program 3.10: *mqam_detector.m*: Coherent detection of MQAM signaling points

```matlab
function [dCap]= mqam_detector(M,r)
%Function to detect MQAM modulated symbols
%[dCap]= mqam_detector(M,r) detects the received MQAM signal points
%points - 'r'. M is the modulation level of MQAM
if(((M~=1) && ~mod(floor(log2(M)),2))==0), %M not a even power of 2
    error('Only Square MQAM supported. M must be even power of 2');
end
ref=constructQAM(M); %reference constellation for MQAM
[~,dCap]= iqOptDetector(r,ref); %IQ detection
end
```

3.5.4 *Optimum detector on IQ plane using minimum Euclidean distance*

Two main categories of detection techniques, commonly applied for detecting the digitally modulated data are *coherent detection* and *non-coherent detection*.

In the vector simulation model for the coherent detection , the transmitter and receiver agree on the same reference constellation for modulating and demodulating the information. The modulators described in section 3.4, incorporate the code to generate the reference constellation for the selected modulation type. The same reference constellation should be used if coherent detection is selected as the method of demodulating the received data vector. On the other hand, in the non-coherent detection , the receiver is oblivious to the reference constellation used at the transmitter. The receiver uses methods like envelope detection to demodulate the data. The IQ detection technique is described here as an example of coherent detection. For an example on the non-coherent detection method, refer Section 3.6.2.1 on non-coherent detection of an MFSK signal vector.

In the IQ detection technique - a type of coherent detection, the first step is to compute the pair-wise Euclidean distance between the given two vectors - reference array and the received symbols corrupted with noise. Each symbol in the received symbol vector (represented on a p-dimensional plane) should be compared with every symbol in the reference array. Next, the symbols, from the reference array, that provide the minimum Euclidean distance are returned.

Let $\mathbf{x} = (x_1, x_2, ..., x_p)$ and $\mathbf{y} = (y_1, y_2, ..., y_p)$ be two points in p-dimensional space. The Euclidean distance between them is given by

$$d(\mathbf{x}, \mathbf{y}) = \sqrt{(x_1 - y_1)^2 + (x_2 - y_2)^2 + ... + (x_p - y_p)^2} \qquad (3.12)$$

The pair-wise Euclidean distance between two sets of vectors, say $\mathbf{x}$ and $\mathbf{y}$, on a p-dimensional space, can be computed using the vectorized Matlab code shown next. Here, $\mathbf{x}$ is a matrix of dimension $m \times p$ and $\mathbf{y}$ is of dimension $n \times p$. The code computes the pair-wise Euclidean distance of each point in the matrix $\mathbf{x}$ against all ideal signal points in the matrix $\mathbf{y}$. It then returns the ideal signaling points from matrix $\mathbf{y}$ that provides the minimum Euclidean distance. Since the vectorized implementation is devoid of nested for-loops, the given vectorized code is significantly faster for larger input matrices. The given code is very generic in the sense that it can be easily reused to implement optimum coherent receivers for any N-dimensional digital modulation technique.

Program 3.11: *minEuclideanDistance.m*: Computing minimum Euclidean distance - a vectorized code

```
1   function [idealPoints,indices]= minEuclideanDistance(x,y)
2   %function to compute the pairwise minimum Distance between two
3   %vectors x and y in p-dimensional signal space and select the
4   %vectors in y that provides the minimum distances.
5   % x - a matrix of size mxp
6   % y - a matrix of size nxp. This acts as a reference against
7   % which each point in x is compared.
8   % idealPoints - contain the decoded vector
9   % indices - indices of the ideal points in reference matrix y
10  [m,p1] = size(x);
11  [n,p2] = size(y);
12
13  if p1~=p2
14      error('Dimension Mismatch: x and y must have same dimension')
15  end
16
17  X = sum(x.*x,2);
18  Y = sum(y.*y,2)';
```

```
19   d = X(:,ones(1,n)) + Y(ones(1,m),:) - 2*x*y';%Squared Euclidean Dist.
20   [~,indices]=min(d,[],2); %Find the minimum value along DIM=2
21   idealPoints=y(indices,:);
22   indices=indices.';
23   end
```

As an example, let's reuse the above snippet of code for implementing a generic optimum detector on IQ plane. This optimum detector accepts two inputs: `received` and `ref`. The vector `received` stands for the sequence of symbols received by the receiver, represented on the complex IQ plane. The vector `ref` stands for the ideal constellation points represented in complex IQ plane. The optimum detector computes the pair-wise Euclidean distance of each point in the received vector against every point in the `ref` vector. It then returns the decoded symbols that provide the minimum Euclidean distance. This optimum detector can used for IQ modulation techniques like M-PSK, M-QAM, M-PAM and the multidimensional MFSK signaling scheme.

Program 3.12: *iqOptDetector.m*: Optimum Detector on IQ plane using minimum Euclidean distance

```
1    function [idealPoints,indices]= iqOptDetector(received,ref)
2    %Optimum Detector for 2-dim. signals (MQAM,MPSK,MPAM) in IQ Plane
3    %received - vector of form I+jQ
4    %ref - reference constellation of form I+jQ
5    %Note: MPAM/BPSK are one dim. modulations. The same function can be
6    %applied for these modulations since quadrature is zero (Q=0).
7    x=[real(received); imag(received)]';%received vec. in cartesian form
8    y=[real(ref); imag(ref)]';%reference vec. in cartesian form
9    [idealPoints,indices]= minEuclideanDistance(x,y);
10   end
```

3.6 M-ary FSK modulation and detection

M-ary FSK modulation belongs to a broader class of orthogonal modulation, where a transmitted signal corresponding to a source symbol, is drawn from a set of M waveforms whose carrier frequencies are orthogonal to each other. M is the size of the alphabet such that each symbol represents $k = log_2 M$ bits from the information source. When $M = 2$, the format shrinks to BFSK modulation.

3.6.1 Modulator for M orthogonal signals

The waveform simulation technique for BFSK was discussed in Section 2.12, where the orthogonality criteria for coherent and non-coherent detection was also derived. The same concept and the simulation technique can be extended for M-ary FSK modulation and detection. However, waveform simulation of M-ary FSK is elaborate and consumes more system memory during run time. A simpler technique to simulate M-ary FSK is to view it in terms of signal space representation. To illustrate this concept we will first see how to represent a BFSK signal using signal space representation and then extend it to represent the broader class of M-ary FSK scheme.

Consider the BFSK modulation scheme, the two waveforms are

$$S_{BFSK}(t) = \begin{cases} s_1(t) = \sqrt{\frac{2E_b}{T_b}} cos\left[2\pi f_1 t + \phi_1\right] & 0 \leq t \leq T_b, \ for \ binary \ 1 \\ s_2(t) = \sqrt{\frac{2E_b}{T_b}} cos\left[2\pi f_2 t + \phi_2\right] & 0 \leq t \leq T_b, \ for \ binary \ 0 \end{cases}$$

Here, ϕ_1 and ϕ_2 are the unknown random phases of the orthogonal carriers that could take any value from the uniform distribution in the interval $[0, 2\pi)$. If $\phi_1 = \phi_2$, then the modulation is called *coherent* BFSK, otherwise it is called *non-coherent* BFSK. Rewriting equation 3.13,

$$s_1(t) = \sqrt{\frac{2E_b}{T_b}} cos(2\pi f_1 t) cos(\phi_1) - \sqrt{\frac{2E_b}{T_b}} sin(2\pi f_1 t) sin(\phi_1)$$

$$s_2(t) = \sqrt{\frac{2E_b}{T_b}} cos(2\pi f_2 t) cos(\phi_2) - \sqrt{\frac{2E_b}{T_b}} sin(2\pi f_2 t) sin(\phi_2) \qquad (3.13)$$

Choosing four basis functions,

$$\psi_{1c}(t) = \sqrt{\frac{2}{T_b}} cos(2\pi f_1 t) \ , \psi_{1s}(t) = -\sqrt{\frac{2}{T_b}} sin(2\pi f_1 t)$$

$$\psi_{2c}(t) = \sqrt{\frac{2}{T_b}} cos(2\pi f_2 t) \ , \psi_{2s}(t) = -\sqrt{\frac{2}{T_b}} sin(2\pi f_2 t) \qquad (3.14)$$

The BFSK waveforms can be expressed in complex notation and the signal set can be conveniently represented as a matrix.

$$\mathbf{s} = \begin{bmatrix} \mathbf{s}_1 \\ \mathbf{s}_2 \end{bmatrix} = \begin{bmatrix} \sqrt{E_b} cos(\phi_1) + j\sqrt{E_b} sin(\phi_1) & 0 \\ 0 & \sqrt{E_b} cos(\phi_2) + j\sqrt{E_b} sin(\phi_2) \end{bmatrix}$$

$$= \begin{bmatrix} \sqrt{E_b} e^{j\phi_1} & 0 \\ 0 & \sqrt{E_b} e^{j\phi_2} \end{bmatrix} \qquad (3.15)$$

For BFSK modulation, if the source bit $a[k] = 0$, then the vector s_1 will be transmitted and if the source bit $a[k] = 1$, the vector s_2 will be transmitted.

Extending the equation 3.15 to M orthogonal frequencies and when $\phi_1 \neq \phi_2 \neq ... \neq \phi_M$, the *noncoherent* M-ary FSK set is represented as

$$\mathbf{s} = \begin{bmatrix} \mathbf{s}_1 \\ \mathbf{s}_2 \\ \vdots \\ \mathbf{s}_M \end{bmatrix} = \begin{bmatrix} \sqrt{E_b} e^{j\phi_1} & 0 & \cdots & 0 \\ 0 & \sqrt{E_b} e^{j\phi_2} & \cdots & 0 \\ \vdots & \vdots & \ddots & \vdots \\ 0 & 0 & \cdots & \sqrt{E_b} e^{j\phi_M} \end{bmatrix} \qquad (3.16)$$

For *coherent* M-ary FSK, $\phi_1 = \phi_2 = ... = \phi_M = 0$.

A function implementing an MFSK modulator, that is capable of choosing between coherent and noncoherent modulation is given here.

Program 3.13: *mfsk_modulator.m*: MFSK modulator for coherent and noncoherent modulation schemes

```
1  function [s,ref]= mfsk_modulator(M,d,COHERENCE)
2  %Function to MFSK modulate the vector of data symbols - d
3  %[s,ref]=mfsk_modulator(M,d,COHERENCE) modulates the symbols defined
4  %by the vector d using MFSK modulation, where M specifies the order
5  %of M-FSK modulation and the vector d contains symbols whose values
6  %in the range 1:M.
```

```
7    %The parameter 'COHERENCE' = 'COHERENT' or 'NONCOHERENT' specifies
8    %the type of MFSK modulation/detection. The output s is the
9    %modulated output and ref represents the reference constellation
10   %that can be used during coherent demodulation.
11   if strcmpi(COHERENCE,'coherent'),
12       phi= zeros(1,M); %phase=0 for coherent detection
13       ref = complex(diag(exp(1i*phi)));%force complex data type
14       s = complex(ref(d,:)); %force complex type, since imag part is 0
15   else
16       phi = 2*pi*rand(1,M);%M random phases in the (0,2pi)
17       ref = diag(exp(1i*phi));
18       s = ref(d,:);
19   end
20   end
```

3.6.2 M-FSK detection

There are two types of detection techniques applicable to MFSK detection: coherent and non-coherent detection. The following function implements the coherent and noncoherent detection of MFSK signal. The theory behind these detection techniques are described next.

Program 3.14: *mfsk_detector.m*: MFSK detector using coherent and noncoherent detection schemes

```
1    function [dCap]= mfsk_detector(M,r,COHERENCE)
2    %Function to detect MFSK modulated symbols
3    %[dCap]= mfsk_detector(M,r,COHERENCE) detects the received MFSK
4    %signal points - 'r'. M is the modulation level of MFSK.
5    %'COHERENCE' = 'COHERENT' or 'NONCOHERENT' specifies the type
6    %of MFSK detection. The output dCap is the detected symbols.
7    if strcmpi(COHERENCE,'coherent'),
8      phi= zeros(1,M); %phase=0 for coherent detection
9      ref = complex(diag(exp(1i*phi)));%reference constellation
10     [~,dCap]= minEuclideanDistance(real(r),ref);%coherent detection
11   else %for noncoherent MFSK
12     [~,dCap]= max(abs(r),[],2);%envelope detection
13     dCap = dCap.';
14   end
15   end
```

3.6.2.1 Non-coherent detection of MFSK

With reference to equation 3.16, if the MFSK modulated symbol vector s_1 is transmitted at an instance, the received signal vector after passing through an AWGN channel is represented as

$$r = \left(\sqrt{E_b} e^{j\phi_1} + n_1, n_2, \cdots, n_M \right) \tag{3.17}$$

where $n_1, n_2, ..., n_M$ are zero-mean statistically independent Gaussian random variables with equal variance $\sigma^2 = N_0/2$. The detection at receiver can be made simple by employing an *envelope detector*, that chooses the ideal transmitted symbol corresponding to the maximum envelope value. This eliminates the need for the carrier phase reference (the term $e^{j\phi}$) at the receiver and hence the technique is called *non-coherent* detection.

3.6.2.2 Coherent detection of MFSK

For coherent equal-energy orthogonal signaling (coherent M-FSK), set $\phi = 0$ in equation 3.16. The signal set for coherent signaling is

$$\mathbf{s} = \begin{bmatrix} \mathbf{s}_1 \\ \mathbf{s}_2 \\ \vdots \\ \mathbf{s}_M \end{bmatrix} = \begin{bmatrix} \sqrt{E_b} & 0 & \cdots & 0 \\ 0 & \sqrt{E_b} & \cdots & 0 \\ \vdots & \vdots & \ddots & \vdots \\ 0 & 0 & \cdots & \sqrt{E_b} \end{bmatrix} \tag{3.18}$$

The optimum receiver for AWGN channel computes the pairwise Euclidean distances between each received symbol and the reference signal set as described in section 3.5.4. The symbols from the reference constellation that provides the minimum Euclidean distance is chosen as the detected symbol. Correlation metric can also be used instead of Euclidean distance metric. In such a case, each received symbol is correlated with each of the vectors in the reference array and the vector that provides the maximum correlation can be chosen.

Note that a coherently modulated MFSK signal can be detected using both coherent and non-coherent detection schemes. However, a non-coherently modulated MFSK can only be detected using noncoherent detection. The figure 3.7 captures the simulation model that can be for MFSK modulation and demodulation over an AWGN channel. For a complete performance simulation model, refer the next chapter.

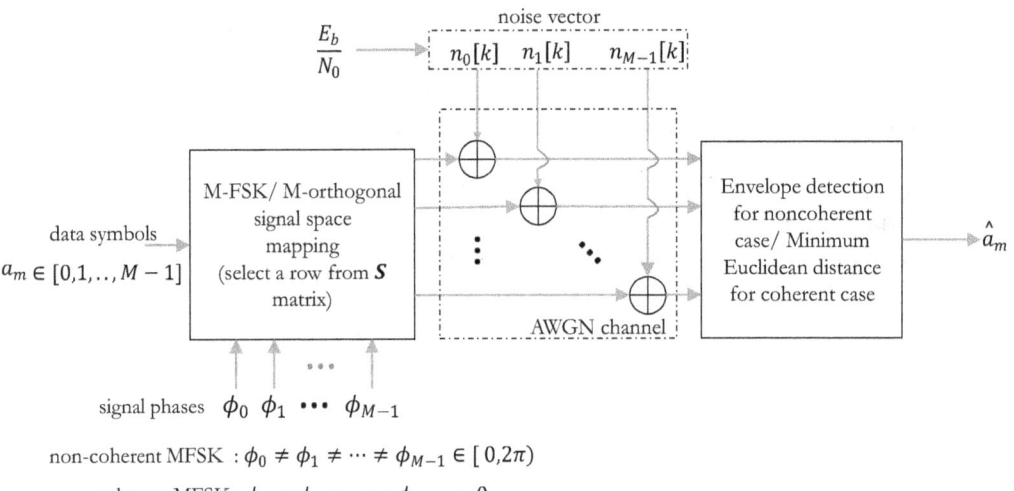

Fig. 3.7: Simulation model for coherent and noncoherently detected MFSK

References

1. Proakis J.G, *Digital Communications*, fifth edition, New York, McGraw–Hill, 2008
2. Jiaqi Zhao et al., *Investigation on performance of special-shaped 8-quadrature amplitude modulation constellations applied in visible light communication*, Photon. Res. 4, 249-256 (2016)

Chapter 4
Performance of Digital Modulations over Wireless Channels

Abstract The focus of this chapter is to develop code for performance simulation of various digital modulation techniques over AWGN and flat fading channels. Complex baseband models, developed in chapter 3, are best suited for performance simulation since they consume less computer memory and yield results quickly. Theoretical performance symbol error rates for AWGN and flat fading channels are also provided in this chapter and that can be used as a benchmark to verify the implemented simulation models

4.1 AWGN channel

In this section, the relationship between SNR-per-bit (E_b/N_0) and SNR-per-symbol (E_s/N_0) are defined with respect to M-ary signaling schemes. Then the complex baseband model for an AWGN channel is discussed, followed by the theoretical error rates of various modulations over the AWGN channel. Finally, the complex baseband models for digital modulators and detectors developed in chapter 3, are incorporated to build a complete communication system model.

4.1.1 Signal to noise ratio (SNR) definitions

Assuming a channel of bandwidth B, received signal power P_r and the power spectral density (PSD) of noise $N_0/2$, the signal to noise ratio (SNR) is given by

$$\gamma = \frac{P_r}{N_0 B} \tag{4.1}$$

Let a signal's energy-per-bit is denoted as E_b and the energy-per-symbol as E_s, then $\gamma_b = E_b/N_0$ and $\gamma_s = E_s/N_0$ are the SNR-per-bit and the SNR-per-symbol respectively. For uncoded M-ary signaling scheme with $k = log_2(M)$ bits per symbol, the signal energy per modulated symbol is given by

$$E_s = k.E_b \tag{4.2}$$

The SNR per symbol is given by

$$\gamma_s = \frac{E_s}{N_0} = k.\frac{E_b}{N_0} = k.\gamma_b \tag{4.3}$$

4.1.2 AWGN channel model

In order to simulate a specific SNR point in performance simulations, the modulated signal from the transmitter needs to be added with random noise of specific strength. The strength of the generated noise depends on the desired SNR level which usually is an input in such simulations. In practice, SNRs are specified in *dB*. Given a specific SNR point for simulation, let's see how we can simulate an AWGN channel that adds correct level of white noise to the transmitted symbols.

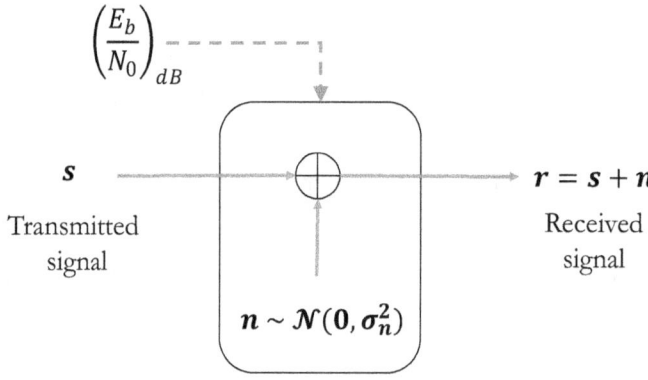

Fig. 4.1: AWGN noise model - computes and adds white Gaussian noise vector for a given SNR value

Consider the AWGN channel model given in Figure 4.1. Given a specific SNR point to simulate, we wish to generate a white Gaussian noise vector of appropriate strength and add it to the incoming signal. The method described can be applied for both waveform simulations and the complex baseband simulations. In following text, the term SNR (γ) refers to $\gamma_b = E_b/N_0$ when the modulation is of binary type (example: BPSK). For multilevel modulations such as QPSK and MQAM, the term SNR refers to $\gamma_s = E_s/N_0$.

1. Assume, **s** is a vector that represents the transmitted signal. We wish to generate a vector **r** that represents the signal after passing through the AWGN channel. The amount of noise added by the AWGN channel is controlled by the given SNR - γ.
2. For waveform simulation (see chapter 2) model, let the given oversampling ratio is denoted as L. On the other hand, if you are using the complex baseband models (see chapter 3), set $L = 1$.
3. Let N denotes the length of the vector **s**. The signal power for the vector **s** can be measured as,

$$P = L \times \frac{1}{N} \sum_{i=0}^{N-1} |s_i|^2 \tag{4.4}$$

4. The required power spectral density of the noise vector **n** is computed as

$$N_0 = \frac{signal\ power}{signal-to-noise-ratio} = \frac{P}{\gamma} \tag{4.5}$$

5. Assuming complex IQ plane for all the digital modulations, the required noise variance (noise power) for generating Gaussian random noise is given by

$$\sigma^2 = \frac{N_0}{2} \tag{4.6}$$

6. Finally, generate the noise vector **n** drawn from normal distribution with mean set to zero and the standard deviation computed from the equation 4.6 given above.

$$\mathbf{n} = \begin{cases} \sigma \times randn(size(\mathbf{s})) & \text{if } \mathbf{s} \text{ is real} \\ \sigma \times \left[randn(size(\mathbf{s})) + j * randn(size(\mathbf{s})) \right] & \text{if } \mathbf{s} \text{ is complex} \end{cases} \qquad (4.7)$$

7. Finally add the generated noise vector to the signal **s**

$$r = \mathbf{s} + \mathbf{n} \qquad (4.8)$$

The following custom function written in Matlab, can be used for adding AWGN noise to an incoming signal. It can be used in waveform simulation as well as complex baseband simulation models.

Program 4.1: *add_awgn_noise.m*: A custom function to add AWGN noise to a signal vector

```
function [r,n,N0] = add_awgn_noise(s,SNRdB,L)
%Function to add AWGN to the given signal
%[r,n,N0]= add_awgn_noise(s,SNRdB) adds AWGN noise vector to signal
%'s' to generate a %resulting signal vector 'r' of specified SNR
%in dB. It also returns the noise vector 'n' that is added to the
%signal 's' and the spectral density N0 of noise added
%
%[r,n,N0]= add_awgn_noise(s,SNRdB,L) adds AWGN noise vector to
%signal 's' to generate a resulting signal vector 'r' of specified
%SNR in dB. The parameter 'L' specifies the oversampling ratio used
%in the system (for waveform simulation). It also returns the noise
%vector 'n' that is added to the signal 's' and the spectral
%density N0 of noise added
s_temp=s;
if iscolumn(s), s=s.'; end; %to return the result in same dim as 's'
gamma = 10^(SNRdB/10); %SNR to linear scale
if nargin==2, L=1; end %if third argument is not given, set it to 1

if isvector(s),
    P=L*sum(abs(s).^2)/length(s);%Actual power in the vector
else %for multi-dimensional signals like MFSK
    P=L*sum(sum(abs(s).^2))/length(s); %if s is a matrix [MxN]
end

N0=P/gamma; %Find the noise spectral density
if(isreal(s)),
    n = sqrt(N0/2)*randn(size(s));%computed noise
else
    n = sqrt(N0/2)*(randn(size(s))+1i*randn(size(s)));%computed noise
end
r = s + n; %received signal
if iscolumn(s_temp), r=r.'; end;%return r in original format as s
end
```

4.1.3 Theoretical symbol error rates

Denoting the Symbol Error Rate (SER) as P_s, SNR-per-bit as $\gamma_b = E_b/N_0$ and SNR-per-symbol as $\gamma_s = E_s/N_0$, the performance error rates for various modulation schemes over AWGN channel are listed in Table 4.1.

Table 4.1: Theoretical Symbol Error rates for various digital modulation schemes

Modulation	Symbol Error Rate (P_s)
MPAM	$2\left(1 - \dfrac{1}{M}\right) Q\left(\sqrt{\dfrac{6}{M^2 - 1}\gamma_s}\right)$
BPSK	$Q\left(\sqrt{2\gamma_b}\right)$
QPSK	$2Q\left(\sqrt{2\gamma_b}\right) - Q^2\left(\sqrt{2\gamma_b}\right)$
MPSK ($M > 4$)	$2Q\left[\sin\left(\dfrac{\pi}{M}\right)\sqrt{2\gamma_s}\right]$
MQAM	$1 - \left[1 - 2\left(1 - \dfrac{1}{\sqrt{M}}\right)Q\left(\sqrt{\dfrac{3\gamma_s}{(M-1)}}\right)\right]^2$
MFSK (noncoherent)	$\displaystyle\sum_{i=1}^{M-1}\dfrac{(-1)^{i+1}}{i+1}\binom{M-1}{i}\exp\left(-\dfrac{i}{i+1}\gamma_s\right)$
MFSK (coherent)	$1 - \displaystyle\int_{-\infty}^{\infty}\left[Q\left(-q - \sqrt{2\gamma_s}\right)\right]^{M-1}\dfrac{1}{\sqrt{2\pi}}\exp\left(-\dfrac{q^2}{2}\right)dq$

The theoretical symbol error rates are coded as a reusable function (shown next). In this implementation, $erfc$ function is used instead of the Q function shown in the Table 4.1. The following equation describes the relationship between the $erfc$ function and the Q function.

$$Q(x) = \frac{1}{2}erfc\left(\frac{x}{\sqrt{2}}\right) \tag{4.9}$$

Program 4.2: *ser_awgn.m*: Theoretical SERs for various modulation schemes over AWGN channel

```
function [SER] = ser_awgn(EbN0dB,MOD_TYPE,M,COHERENCE)
%Theoretical Symbol Error Rate for various modulations over AWGN
%EbN0dB - list of SNR per bit values
%MOD_TYPE - 'BPSK','PSK','QAM','PAM','FSK'
%M - Modulation level for the chosen modulation
%    - For PSK,PAM,FSK M can be any power of 2
%    - For QAM M must be even power of 2 (square QAM only)
%Parameter COHERENCE is only applicable for FSK modulation
%COHERENCE = 'coherent' for coherent FSK detection
%          = 'noncoherent' for noncoherent FSK detection
gamma_b = 10.^(EbN0dB/10); %SNR per bit in linear scale
gamma_s = log2(M)*gamma_b; %SNR per symbol in linear scale
SER = zeros(size(EbN0dB));

```

```matlab
15  switch lower(MOD_TYPE)
16      case 'bpsk'
17          SER=0.5*erfc(sqrt(gamma_b));
18      case {'psk','mpsk'}
19          if M==2, %for BPSK
20              SER=0.5*erfc(sqrt(gamma_b));
21          else
22              if M==4, %for QPSK
23                  Q=0.5*erfc(sqrt(gamma_b)); SER=2*Q-Q.^2;
24              else %for other higher order M-ary PSK
25                  SER=erfc(sqrt(gamma_s)*sin(pi/M));
26              end
27          end
28      case {'qam','mqam'}
29          SER = 1-(1-(1-1/sqrt(M))*erfc(sqrt(3/2*gamma_s/(M-1)))).^2;
30      case {'fsk','mfsk'}
31          if strcmpi(COHERENCE,'coherent'),
32              for ii=1:length(gamma_s),
33                  fun = @(q) (0.5*erfc((-q - sqrt(2.*gamma_s(ii)))/sqrt(2))).^(M-1).*
                          1/sqrt(2*pi).*exp(-q.^2/2);
34                  SER(ii) = 1-integral(fun,-inf,inf);
35              end
36          else %Default compute for noncoherent
37              for jj=1:length(gamma_s),
38                  summ=0;
39                  for i=1:M-1,
40                      n=M-1; r=i; %for nCr formula
41                      summ=summ+(-1).^(i+1)./(i+1).*prod((n-r+1:n)./(1:r)).*exp(-i./(i
                              +1).*gamma_s(jj));
42                  end
43                  SER(jj)=summ; %Theoretical SER for non-coherent detection
44              end
45          end
46      case {'pam','mpam'}
47          SER=2*(1-1/M)*0.5*erfc(sqrt(3*gamma_s/(M^2-1)));
48      otherwise
49          display 'ser_awgn.m: Invalid modulation (MOD_TYPE) selected'
50  end
51  end
```

4.1.4 Unified simulation model for performance simulation

In chapter 3, the code implementation for complex baseband models for various digital modulators and demodulator are given. The AWGN channel model is given section 4.1.2. Using these models, we can create a unified simulation for code for simulating the performance of various modulation techniques over AWGN channel.

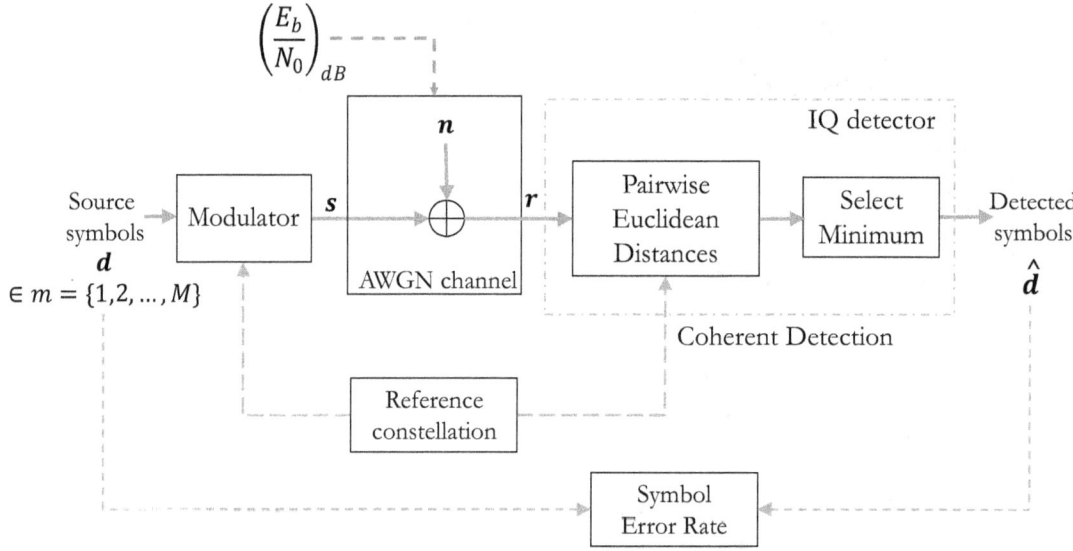

Fig. 4.2: Performance simulation model illustrated for a receiver employing coherent detection

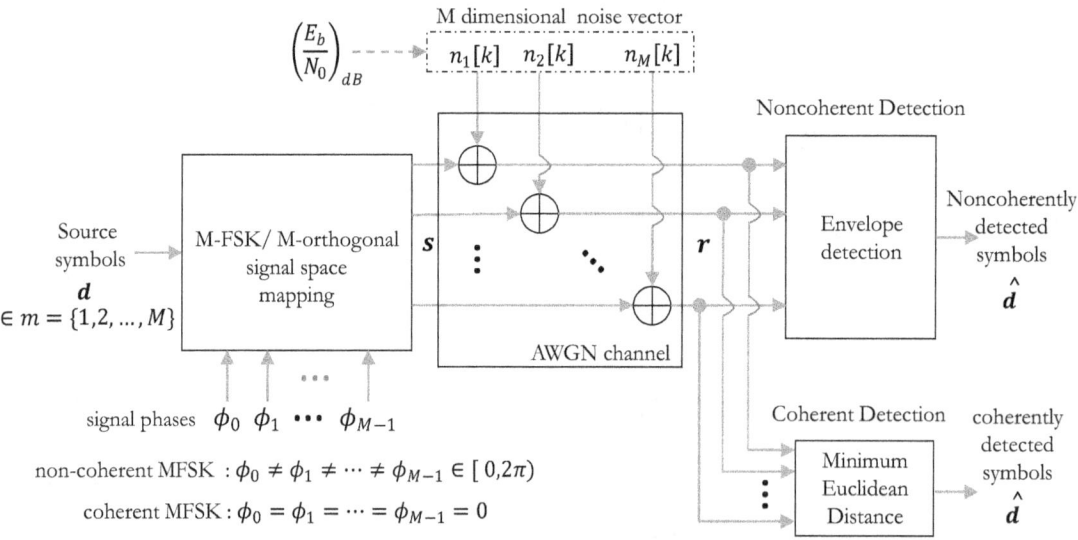

Fig. 4.3: Performance simulation model illustrated for MFSK modulation over AWGN channel

The complete simulation model for performance simulation over AWGN channel is given in Figure 4.2. The figure is illustrated for a *coherent* communication system model (applicable for MPSK/MQAM/M-PAM/MFSK modulations)

Figure 4.3 illustrates the simulation model for multi-level signaling like MFSK simulation. As discussed in section 3.6 of chapter 3, a coherently modulated MFSK signal can be detected using both coherent detection and noncoherent detection technique. However, if the MFSK modulator employs noncoherent signal generation, the receiver can only employ noncoherent detection technique.

The Matlab code implementing the aforementioned simulation model is given next. We have employed an unified approach to simulate the performance of any of the given modulation technique - MPSK, MQAM, MPAM or MFSK. To simulate the performance for a given modulation, the user just needs to set the MOD_TYPE to 'PAM' or 'PSK' or 'QAM' or 'FSK'. If FSK is chosen a the modulation type, the user needs to specify if the receiver employs 'COHERENT' or 'NONCOHERENT' detection technique, by setting the COHERENCE variable in the code.

The simulation code will automatically choose the selected modulation type, performs Monte Carlo simulation, computes symbol error rates and plots them against the theoretical symbol error rates. The simulated performance results obtained for various modulations are shown in the Figure 4.4.

Program 4.3: *perf_over_awgn.m*: Performance of various modulations over AWGN channel

```
 1  %Eb/N0 Vs SER for PSK/QAM/PAM/FSK over AWGN (complex baseband model)
 2
 3  clearvars; clc;
 4  %---------Input Fields-----------------------
 5  nSym=10^6;%Number of symbols to transmit
 6  EbN0dB = -4:2:14; % Eb/N0 range in dB for simulation
 7  MOD_TYPE='FSK'; %Set 'PSK' or 'QAM' or 'PAM' or 'FSK'
 8  arrayOfM=[2,4,8,16,32]; %array of M values to simulate
 9  %arrayOfM=[4,16,64,256]; %uncomment this line if MOD_TYPE='QAM'
10  COHERENCE = 'noncoherent';%'coherent'/'noncoherent'-only for FSK
11
12  plotColor =['b','g','r','c','m','k']; p=1; %plot colors
13  legendString = cell(1,length(arrayOfM)*2); %for legend entries
14
15  for M = arrayOfM
16      %-----Initialization of various parameters----
17      k=log2(M);
18      EsN0dB = 10*log10(k)+EbN0dB; %EsN0dB calculation
19      SER_sim = zeros(1,length(EbN0dB));%simulated Symbol error rates
20
21      d=ceil(M.*rand(1,nSym));%uniform random symbols from 1:M
22      s=modulate(MOD_TYPE,M,d,COHERENCE);%(Refer Chapter 3)
23
24      for i=1:length(EsN0dB),
25          r = add_awgn_noise(s,EsN0dB(i));%add AWGN noise
26
27          dCap = demodulate(MOD_TYPE,M,r,COHERENCE);%(Refer Chapter 3)
28          SER_sim(i) = sum((d~=dCap))/nSym;%SER computation
29      end
30
31      SER_theory = ser_awgn(EbN0dB,MOD_TYPE,M,COHERENCE);%theory SER
32      semilogy(EbN0dB,SER_sim,[plotColor(p) '*']);
```

```
33    hold on;
34    semilogy(EbN0dB,SER_theory,plotColor(p));
35    legendString{2*p-1}=['Sim ',num2str(M),'-',MOD_TYPE];
36    legendString{2*p}=['Theory ',num2str(M),'-',MOD_TYPE];
37    p=p+1;
38  end
39  legend(legendString);
40  xlabel('Eb/N0(dB)');
41  ylabel('SER (Ps)');
42  title(['Probability of Symbol Error for M-',MOD_TYPE,' over AWGN']);
```

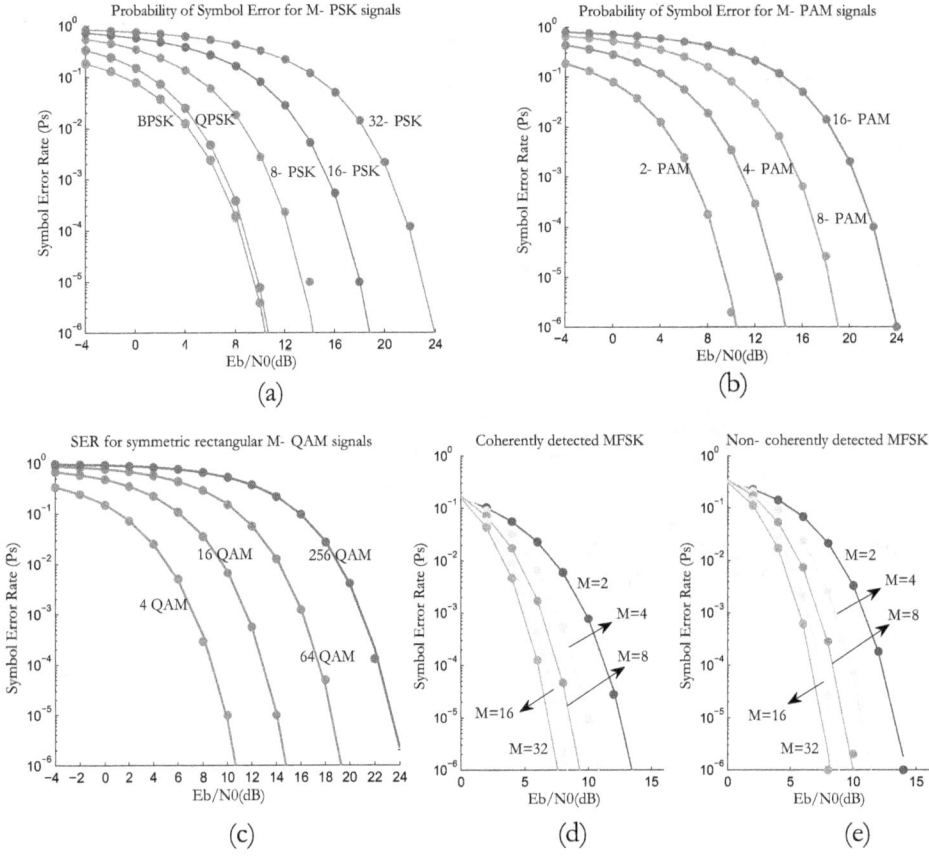

Fig. 4.4: Performance of modulations over AWGN channel : (a) MPSK , (b) MPAM, (c) MQAM, (d) coherently detected MFSK (e) noncoherently detected MFSK

4.2 Fading channels

With respect to the frequency domain characteristics, the fading channels can be classified into *frequency selective* and *flat fading* channel . The channel fading can be modeled with different statistics like Rayleigh, Rician, Nakagami fading. The simulation models in this section are focused on obtaining the simulated performance of various modulations over Rayleigh flat fading and Rician flat fading channels. Modeling of simulating frequency selective fading channel is beyond the scope of this book.

4.2.1 Linear time invariant channel model and FIR filters

The most significant feature of a real world channel is that the channel does not immediately respond to the input. Physically, this indicates some sort of inertia built into the channel/medium, that takes some time to respond. As a consequence, it may introduce distortion effects like *inter-symbol interference* (ISI) at the channel output. Such effects are best studied with the *Linear Time Invariant* (LTI) channel model, given in Figure 4.5.

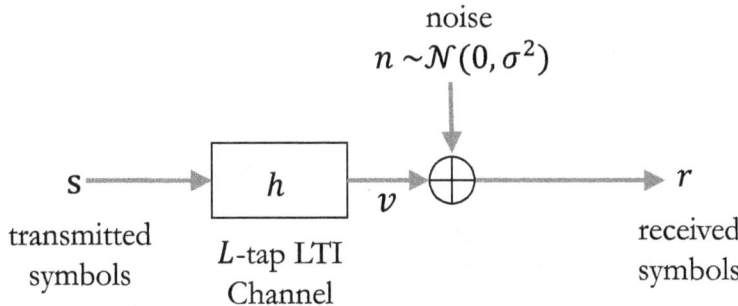

Fig. 4.5: LTI channel model

In this model, the channel response to any input depends only on the *Channel Impulse Response* (CIR) function of the channel. The CIR is usually defined for finite length L as $h = [h_0, h_1, h_2, ..., h_{L-1}]$ where h_0 is the CIR at symbol sampling instant $0T_{sym}$ and h_{L-1} is the CIR at symbol sampling instant $(L-1)T_{sym}$. Such a channel can be modeled as a *Tapped Delay Line* (TDL) filter, otherwise called *Finite Impulse Response* (FIR) filter. Here, we only consider the CIR at symbol sampling instances. It is well known that the output of such a channel (r) is given as the *linear convolution* of the input symbols (s) and the CIR (h) at symbol sampling instances. In addition, channel noise in the form of AWGN can also be included the model. Therefore, the resulting vector of from the entire channel model is given as

$$r = \mathbf{h} * s + n \tag{4.10}$$

4.2.2 Simulation model for detection in flat fading channel

A flat-fading (a.k.a *frequency-non-selective*) channel is modeled with a single tap FIR filter with the tap weights drawn from distributions like Rayleigh, Rician or Nakagami distributions. We will assume *block fading*, which implies that the fading process is approximately constant for a given transmission interval. For

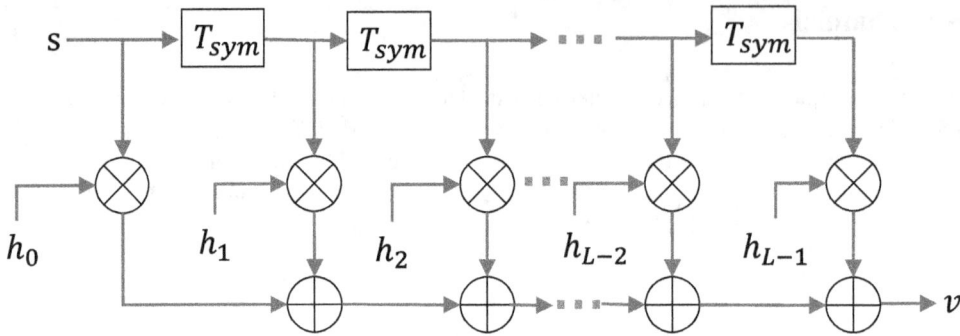

Fig. 4.6: LTI channel viewed as a Tapped Delay Line (TDL) filter

block fading, the random tap coefficient $h[k]$ is a *complex* random variable (not random processes) and for each channel realization, a new set of random values are drawn from Rayleigh or Rician or Nakagami fading according to the type of fading desired.

Simulation models for modulation and detection over a fading channel is shown in Figure 4.7. For a flat fading channel, the output of the channel can be expressed simply as product of time varying channel response and the input signal. Thus equation 4.10 can be simplified as follows for the flat fading channel.

$$r = \mathbf{h}s + n \tag{4.11}$$

Since the channel and noise are modeled as a complex vectors, the detection of s from the received signal is an estimation problem in the complex vector space.

Assuming perfect channel knowledge at the receiver and coherent detection, the receiver shown in Figure 4.7(a) performs *matched filtering*. Matched filtering is effected by multiplying the received signal with the complex conjugate of the channel impulse response h^*. It then scales down the result by $||h||^2$. The resulting *decision vector* is used by the detector block for the estimation of the transmitted vector. In a nutshell, the *sufficient statistic* for the estimation of s from the received signal r is given by (refer equation A.77 in [1])

$$\tilde{y} = \frac{h^*}{||h||^2}r = \frac{h^*}{||h||^2}hs + \frac{h^*}{||h||^2}n = s + \tilde{w} \tag{4.12}$$

Based on this theory, we can construct a simplified model simulating a communication system over flat-fading as in the Figure 4.7(b). The multiplication and division by the factor $|h|$ is a simplification of the combined effect of the channel h, its matched filter h^* and the scaling factor $1/||h||^2$.

To simulate flat fading, the values for the fading variable $\mathbf{h}$ is drawn complex normal distribution

$$\mathbf{h} = (X + jY) \tag{4.13}$$

where X, Y are statistically independent real valued normal random variables.

- If $E[h] = 0$, then $|h|$ is Rayleigh distributed, resulting in a Rayleigh flat fading channel
- If $E[h] \neq 0$, then $|h|$ is Rician distributed, resulting in a Rician flat fading channel with the factor $K = [E[h]]^2/\sigma_h^2$

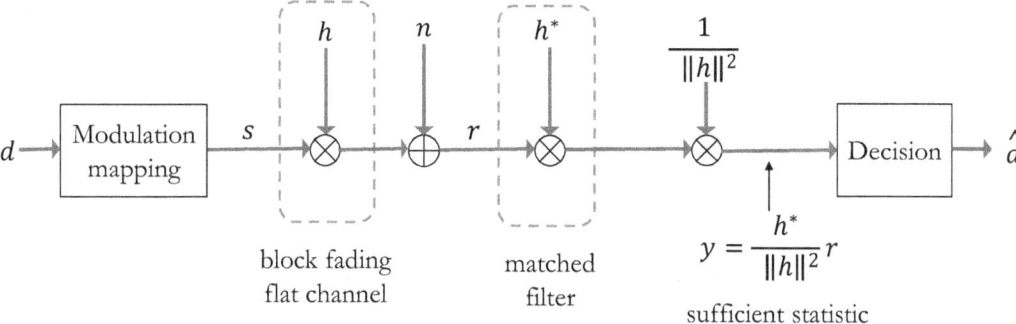

(a) Estimation in additive Gaussian noise in a complex vector space

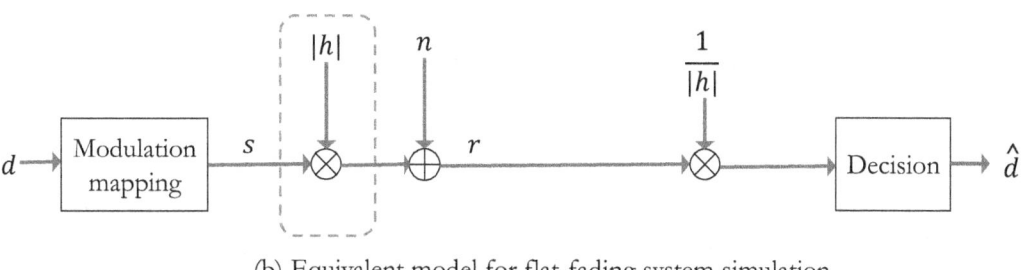

(b) Equivalent model for flat-fading system simulation

Fig. 4.7: Simulation model for modulation and detection over flat fading channel

4.2.3 Rayleigh flat-fading channel

With respect to the simulation model shown in Figure 4.7(b), the samples for the Rayleigh flat-fading samples are drawn from the following random variable

$$\mathbf{h} = ||X + jY|| \tag{4.14}$$

where $X \sim \mathcal{N}(0, \sigma^2/2)$ and $Y \sim \mathcal{N}(0, \sigma^2/2)$. The average power in the distribution is $P_{av} = \sigma^2$. Therefore to model a channel with $P_{av} = 1$, the normal random variables X and Y should have the standard deviation $\sigma = 1/\sqrt{(2)}$. In Matlab, a Rayleigh flat fading channel for N channel realizations can be simulated by the following code

Program 4.4: Generating channel samples for Rayleigh flat-fading

```
1  h=1/sqrt(2)*(randn(1,N)+1i*randn(1,N));%1 tap complex gaussian filter
2  fading=abs(h); %Rayleigh fading channel samples
```

4.2.3.1 Theoretical Symbol Error Rates

The theoretical average probability of symbol errors over Rayleigh fading channel with AWGN noise can be obtained using Moment Generating Function (MGF) approach [2] [3]. Table 4.2 lists the theoretical symbol error rates for various modulations over Rayleigh fading channel with AWGN noise where the parameter

$\bar{\gamma}_s$ denotes the average SNR-per-symbol (E_s/N_0) and the Moment Generating Function (MGF) for Rayleigh distribution is defined as

$$\mathcal{M}_{\gamma_s}\left(-\frac{g}{sin^2\phi}\right) = \left(1 + \frac{g\bar{\gamma}_s}{sin^2\phi}\right)^{-1} \tag{4.15}$$

Table 4.2: Theoretical Symbol Error rates for various digital modulation schemes

Modulation	Parameter g	Avg Prob. of Symbol error $\bar{P}_s$
BPSK	-	$0.5\left(1 - \sqrt{\frac{\bar{\gamma}_s}{1+\bar{\gamma}_s}}\right)$
MPSK	$sin^2\left(\frac{\pi}{M}\right)$	$\frac{1}{\pi}\int_0^{\frac{(M-1)\pi}{M}} \mathcal{M}_{\gamma_s}\left(-\frac{g}{sin^2\phi}\right)d\phi$
MQAM	$\frac{1.5}{(M-1)}$	$\frac{4}{\pi}\left(1-\frac{1}{\sqrt{M}}\right)\int_0^{\pi/2}\mathcal{M}_{\gamma_s}\left(-\frac{g}{sin^2\phi}\right)d\phi - \frac{4}{\pi}\left(1-\frac{1}{\sqrt{M}}\right)^2\int_0^{\pi/4}\mathcal{M}_{\gamma_s}\left(-\frac{g}{sin^2\phi}\right)d\phi$
MPAM	$\frac{3}{(M^2-1)}$	$\frac{2(M-1)}{M\pi}\int_0^{\pi/2}\mathcal{M}_{\gamma_s}\left(-\frac{g}{sin^2\phi}\right)d\phi$

The following Matlab function computes the theoretical symbol errors for various modulation schemes over a Rayleigh fading channel with AWGN noise.

Program 4.5: *ser_rayleigh.m*: Theoretical symbol error rates over Rayleigh fading channel

```
1  function [ser] = ser_rayleigh(EbN0dB,MOD_TYPE,M)
2  %Compute Theoretical Symbol Error rates for MPSK or MQAM modulations
3  %EbN0dB - list of SNR per bit points
4  %MOD_TYPE - 'MPSK' or 'MQAM'
5  %M - Modulation level for the chosen modulation
6  % - For MPSK M can be any power of 2
7  % - For MQAM M must be even power of 2 (square QAM only)
8  gamma_b = 10.^(EbN0dB/10); %SNR per bit in linear scale
9  gamma_s = log2(M)*gamma_b; %SNR per symbol in linear scale
10 switch lower(MOD_TYPE)
11     case {'bpsk'}
12         ser = 0.5*(1-sqrt(gamma_b/(1+gamma_b)));
13     case {'mpsk','psk'}
14         ser = zeros(size(gamma_s));
15         for i=1:length(gamma_s), %for each SNR point
16             g = sin(pi/M).^2;
17             fun = @(x) 1./(1+(g.*gamma_s(i)./(sin(x).^2))); %MGF
18             ser(i) = (1/pi)*integral(fun,0,pi*(M-1)/M);
19         end
20     case {'mqam','qam'}
21         ser = zeros(size(gamma_s));
22         for i=1:length(gamma_s), %for each SNR point
23             g = 1.5/(M-1);
24             fun = @(x) 1./(1+(g.*gamma_s(i)./(sin(x).^2)));%MGF
25             ser(i) = 4/pi*(1-1/sqrt(M))*integral(fun,0,pi/2)-4/pi*(1-1/sqrt(M))^2*
                   integral(fun,0,pi/4);
26         end
```

```
27   case {'mpam','pam'}
28       ser = zeros(size(gamma_s));
29       for i=1:length(gamma_s), %for each SNR point
30           g = 3/(M^2-1);
31           fun = @(x) 1./(1+(g.*gamma_s(i)./(sin(x).^2)));%MGF
32           ser(i) = 2*(M-1)/(M*pi)*integral(fun,0,pi/2);
33       end
34   end;end
```

4.2.3.2 Simulation code and performance results

In chapter 3, the code implementation for complex baseband models for various digital modulators and demodulator are given. The computation and generation of AWGN noise is given section 4.1.2. Use these models, we can create a unified simulation for code for simulating the performance of various modulation techniques over Rayleigh flat-fading channel the simulation model shown in Figure 4.7(b).

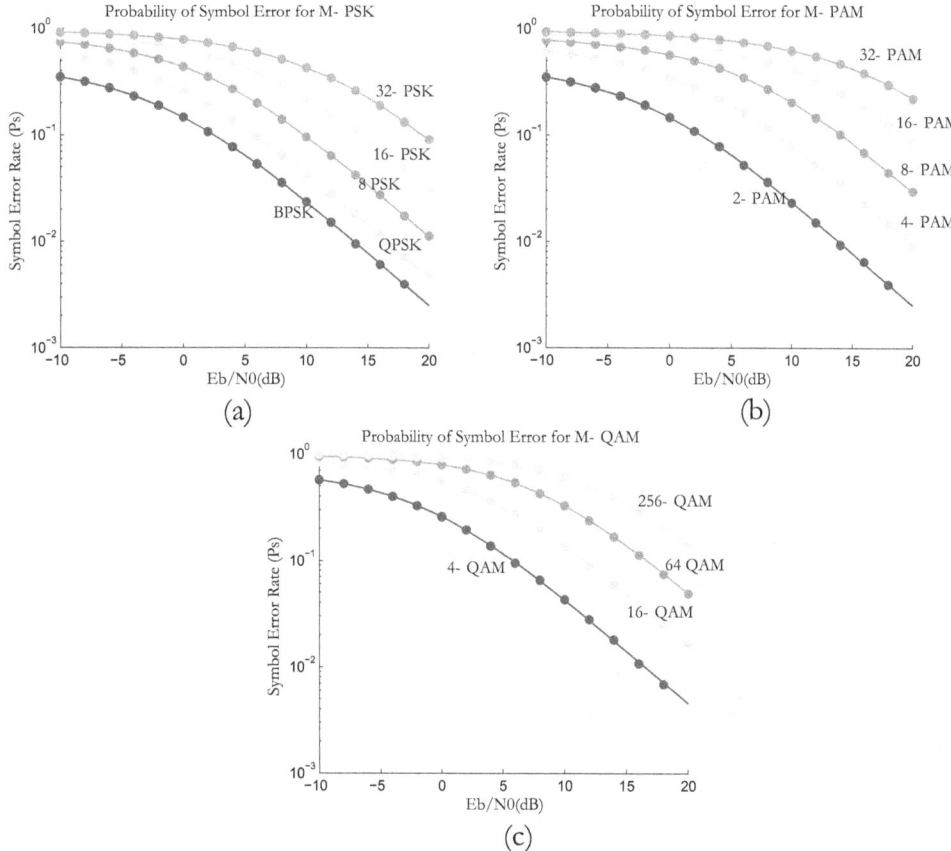

Fig. 4.8: Performance of modulations over Rayleigh flat fading channel : (a) MPSK , (b) MPAM, (c) MQAM

The Matlab code implementing the aforementioned simulation model is given next. An unified approach is employed to simulate the performance of any of the given modulation technique - MPSK, MQAM or MPAM.

To simulate the performance for a given modulation, the user just needs to set the MOD_TYPE to 'PAM' or 'PSK' or 'QAM'.

The simulation code will automatically choose the selected modulation type, performs Monte Carlo simulation, computes symbol error rates and plots them against the theoretical symbol error rates. The simulated performance results obtained for various modulations are shown in the Figure 4.8.

Program 4.6: *perf_over_rayleigh_flat_fading.m*: Performance of modulations over Rayleigh flat fading

```
1   %Performance of PSK/QAM/PAM over Rayleigh flat fading (complex baseband)
2   clearvars; clc;
3   %---------Input Fields-----------------------
4   nSym=10^5;%Number of symbols to transmit
5   EbN0dB = -10:2:20; % Eb/N0 range in dB for simulation
6   MOD_TYPE='QAM'; %Set 'PSK' or 'QAM' or 'PAM' (FSK not supported)
7   arrayOfM=[2,4,8,16,32]; %array of M values to simulate
8   %arrayOfM=[4,16,64,256]; %uncomment this line if MOD_TYPE='QAM'
9
10  plotColor =['b','g','r','c','m','k']; p=1; %plot colors
11  legendString = cell(1,length(arrayOfM)*2); %for legend entries
12
13  for M = arrayOfM
14      %-----Initialization of various parameters----
15      k=log2(M); EsN0dB = 10*log10(k)+EbN0dB; %EsN0dB calculation
16      SER_sim = zeros(1,length(EbN0dB));%simulated Symbol error rates
17
18      d=ceil(M.*rand(1,nSym));%uniform random symbols from 1:M
19      s=modulate(MOD_TYPE,M,d);%(Refer Chapter 3)
20
21      for i=1:length(EsN0dB),
22          h = 1/sqrt(2)*(randn(1,nSym)+1i*randn(1,nSym));%flat Rayleigh
23          hs = abs(h).*s; %flat fading effect on the modulated symbols
24          r = add_awgn_noise(hs,EsN0dB(i));%(Refer 4.1.2)
25
26          %----------------Receiver--------------------
27          y = r./abs(h); %decision vector
28          dCap = demodulate(MOD_TYPE,M,y);%(Refer Chapter 3)
29          SER_sim(i) = sum((d~=dCap))/nSym;%SER computation
30      end
31      SER_theory = ser_rayleigh(EbN0dB,MOD_TYPE,M); %theoretical SER
32      semilogy(EbN0dB,SER_sim,[plotColor(p) '*']); hold on;
33      semilogy(EbN0dB,SER_theory,plotColor(p));
34      legendString{2*p-1}=['Sim ',num2str(M),'-',MOD_TYPE];
35      legendString{2*p}=['Theory ',num2str(M),'-',MOD_TYPE]; p=p+1;
36  end
37  legend(legendString);xlabel('Eb/N0(dB)');ylabel('SER (Ps)');
38  title(['Probability of Symbol Error for M-',MOD_TYPE,' over Rayleigh flat fading
        channel']);
```

4.2.4 Rician flat-fading channel

In wireless environments, a transmitted signal may be subjected to multiple scatterings before arriving at the receiver. This gives rise to random fluctuations in the received signal and this phenomenon is called *fading*. The scattered version of the signal is designated as *Non Line of Sight (NLOS)* component. If the number of NLOS components are sufficiently large, the fading process is approximated as the sum of large number of complex Gaussian process whose probability-density-function follows Rayleigh distribution.

Rayleigh distribution is well suited for the absence of a dominant *Line of Sight (LOS)* path between the transmitter and the receiver. If a line of sight path do exist, the envelope distribution is no longer Rayleigh, but Rician. If there exist a dominant LOS component, the fading process can be represented as the sum of complex exponential and a narrowband complex Gaussian process $g(t)$. If the LOS component arrive at the receiver at an *Angle of Arrival (AoA)* θ, phase ϕ and with the maximum Doppler frequency f_D, the fading process in baseband can be represented as [4]

$$h(t) = \underbrace{\sqrt{\frac{K\Omega}{K+1}}\, e^{(j2\pi f_D cos(\theta)t+\phi)}}_{A:=} + \underbrace{\sqrt{\frac{\Omega}{K+1}}\, g(t)}_{S:=} \tag{4.16}$$

where, K represents the *Rician K factor* given as the ratio of power of LOS component (A^2) to the power of scattered components (S^2) marked in the equation above.

$$K = \frac{A^2}{S^2} \tag{4.17}$$

The received signal power Ω is the sum of power in LOS component and the power in scattered components, given as $\Omega = A^2 + S^2$. The above mentioned fading process is called Rician fading process. The best- and worst-case Rician fading channels are associated with $K = \infty$ and $K = 0$ respectively. A Rician fading channel with $K = \infty$ is a Gaussian channel with a strong LOS path. Rician channel with $K = 0$ represents a Rayleigh channel with no LOS path.

With respect to the simulation model shown in Figure 4.7(b), given a K factor, the samples for the Rician flat-fading samples are drawn from the following random variable

$$\mathbf{h} = ||X + jY|| \tag{4.18}$$

where $\mathbf{X}, \mathbf{Y} \sim \mathcal{N}(\mu, \sigma^2)$ are Gaussian random variables with non-zero mean μ and standard deviation σ as given in [5], [6]

$$\mu = \sqrt{\frac{K}{2(K+1)}} \qquad \sigma = \sqrt{\frac{1}{2(K+1)}} \tag{4.19}$$

Program 4.7: Generating channel samples for Rician flat-fading

```
1  K = 10.^(K_dB/10); %K factor in linear scale
2  mu = sqrt(K/(2*(K+1))); %mean
3  sigma = sqrt(1/(2*(K+1))); %sigma
4  %1 tap complex gaussian filter with Pav=1 and non-zero mean
5  h = (sigma*randn(1,nSym)+mu)+1i*(sigma*randn(1,nSym)+mu);
6  fading = abs(h); %Rician fading channel samples
```

4.2.4.1 Theoretical Symbol Error Rates

The theoretical average probability of symbol errors over Rician flat-fading channel with AWGN noise can be obtained using Moment Generating Function (MGF) approach [2] [3]. Table 4.2 lists the theoretical symbol error rates for various modulations over a fading channel with AWGN noise where the parameter $\bar{\gamma}_s$ denotes the average SNR-per-symbol (E_s/N_0) and the Moment Generating Function (MGF) should correspond to the following equation for the case of Rician fading.

$$\mathscr{M}_{\gamma_s}\left(-\frac{g}{sin^2\phi}\right) = \frac{(1+K)sin^2\phi}{(1+K)sin^2\phi + g\bar{\gamma}_s}exp\left(-\frac{Kg\bar{\gamma}_s}{(1+K)sin^2\phi + g\bar{\gamma}_s}\right) \tag{4.20}$$

The following Matlab function computes the theoretical symbol errors for various modulation schemes over a Rician fading channel with AWGN noise.

Program 4.8: *ser_rician.m*: Theoretical symbol error rates over Rician fading channel

```matlab
function [ser] = ser_rician(EbN0dB,K_dB,MOD_TYPE,M)
%Compute Theoretical Symbol Error rates for MPSK or MQAM modulations
%EbN0dB - list of SNR per bit points
%K_dB - K factor for Rician fading in dB
%MOD_TYPE - 'MPSK' or 'MQAM'
%M - Modulation level for the chosen modulation
% - For MPSK M can be any power of 2
% - For MQAM M must be even power of 2 (square QAM only)
gamma_b - 10.^(FbN0dB/10); %SNR per bit in linear scale
gamma_s = log2(M)*gamma_b; %SNR per symbol in linear scale
K=10^(K_dB/10); %K factor in linear scale

switch lower(MOD_TYPE)
    case {'mpsk','psk'}
        ser = zeros(size(gamma_s));
        for i=1:length(gamma_s), %for each SNR point
            g = sin(pi/M).^2;
            fun = @(x) ((1+K)*sin(x).^2)/((1+K)*sin(x).^2+g*gamma_s(i)).*exp(-K*g*
                gamma_s(i)./((1+K)*sin(x).^2+g*gamma_s(i))); %MGF
            ser(i) = (1/pi)*integral(fun,0,pi*(M-1)/M);
        end
    case {'mqam','qam'}
        ser = zeros(size(gamma_s));
        for i=1:length(gamma_s), %for each SNR point
            g = 1.5/(M-1);
            fun = @(x) ((1+K)*sin(x).^2)/((1+K)*sin(x).^2+g*gamma_s(i)).*exp(-K*g*
                gamma_s(i)./((1+K)*sin(x).^2+g*gamma_s(i))); %MGF
            ser(i) = 4/pi*(1-1/sqrt(M))*integral(fun,0,pi/2)-4/pi*(1-1/sqrt(M))^2*
                integral(fun,0,pi/4);
        end
    case {'mpam','pam'}
        ser = zeros(size(gamma_s));
        for i=1:length(gamma_s), %for each SNR point
            g = 3/(M^2-1);
            fun = @(x) ((1+K)*sin(x).^2)/((1+K)*sin(x).^2+g*gamma_s(i)).*exp(-K*g*
                gamma_s(i)./((1+K)*sin(x).^2+g*gamma_s(i))); %MGF
```

```
33              ser(i) = 2*(M-1)/(M*pi)*integral(fun,0,pi/2);
34         end
35  end
36  end
```

4.2.4.2 Simulation code and performance results

In chapter 3, the code implementation for complex baseband models for various digital modulators and demodulator are given. The computation and generation of AWGN noise is given section 4.1.2. Use these models, we can create a unified simulation for code for simulating the performance of various modulation techniques over Rician flat-fading channel the simulation model shown in Figure 4.7(b).

The Matlab code implementing the aforementioned simulation model is given next. An unified approach is employed to simulate the performance of any of the given modulation technique - MPSK, MQAM or MPAM. To simulate the performance for a given modulation, the user just needs to set the MOD_TYPE to 'PAM' or 'PSK' or 'QAM'.

The simulation code will automatically choose the selected modulation type, performs Monte Carlo simulation, computes symbol error rates and plots them against the theoretical symbol error rate curves. The simulated performance results obtained for various modulations are shown in the Figure 4.9.

Program 4.9: *perf_over_rician_flat_fading.m*: Performance of modulations over Rician flat fading channel

```
1   %Performance of PSK/QAM/PAM over Rician flat fading
2   clearvars; clc;
3   nSym=10^6;%Number of symbols to transmit
4   EbN0dB = 0:2:20; % Eb/N0 range in dB for simulation
5   K_dB = [3,5,10,20]; %array of K factors for Rician fading in dB
6   MOD_TYPE='QAM'; %Set 'PSK' or 'QAM' or 'PAM' (FSK not supported)
7   M=64; %M value for the modulation to simulate
8
9   plotColor =['b','g','r','c','m','k']; p=1; %plot colors
10  legendString = cell(1,length(K_dB)*2); %for legend entries
11
12  for j = 1:length(K_dB),
13      k=log2(M); EsN0dB = 10*log10(k)+EbN0dB; %EsN0dB calculation
14      SER_sim = zeros(1,length(EbN0dB));%simulated Symbol error rates
15
16      d=ceil(M.*rand(1,nSym));%uniform random symbols from 1:M
17      s=modulate(MOD_TYPE,M,d);%(Refer Chapter 3)
18
19      K = 10.^(K_dB(j)/10); %K factor in linear scale
20      mu = sqrt(K/(2*(K+1))); %For Rice fading
21      sigma = sqrt(1/(2*(K+1))); %For Rice fading
22
23      for i=1:length(EsN0dB),
24          h = (sigma*randn(1,nSym)+mu)+1i*(sigma*randn(1,nSym)+mu);
25          display(mean(abs(h).^2));%avg power of the fading samples
26          hs = abs(h).*s; %Rician fading effect on modulated symbols
27          r = add_awgn_noise(hs,EsN0dB(i));%(Refer 4.1.2)
28
```

```
29        %----------------Receiver--------------------
30        y = r./abs(h); %decision vector
31        dCap = demodulate(MOD_TYPE,M,y);%(Refer Chapter 3)
32        SER_sim(i) = sum((d~=dCap))/nSym;%symbol error rate
33      end
34      SER_theory = ser_rician(EbN0dB,K_dB(j),MOD_TYPE,M);%theoretical SER
35      semilogy(EbN0dB,SER_sim,[plotColor(p) '*']); hold on;
36      semilogy(EbN0dB,SER_theory,plotColor(p));
37      legendString{2*p-1}=['Sim K=',num2str(K_dB(j)), ' dB'];
38      legendString{2*p}=['Theory K=',num2str(K_dB(j)), ' dB']; p=p+1;
39    end
40    legend(legendString);xlabel('Eb/N0(dB)');ylabel('SER (Ps)');
41    title(['Probability of Symbol Error for ',num2str(M),'-',MOD_TYPE,' over Rician
          flat fading channel']);
```

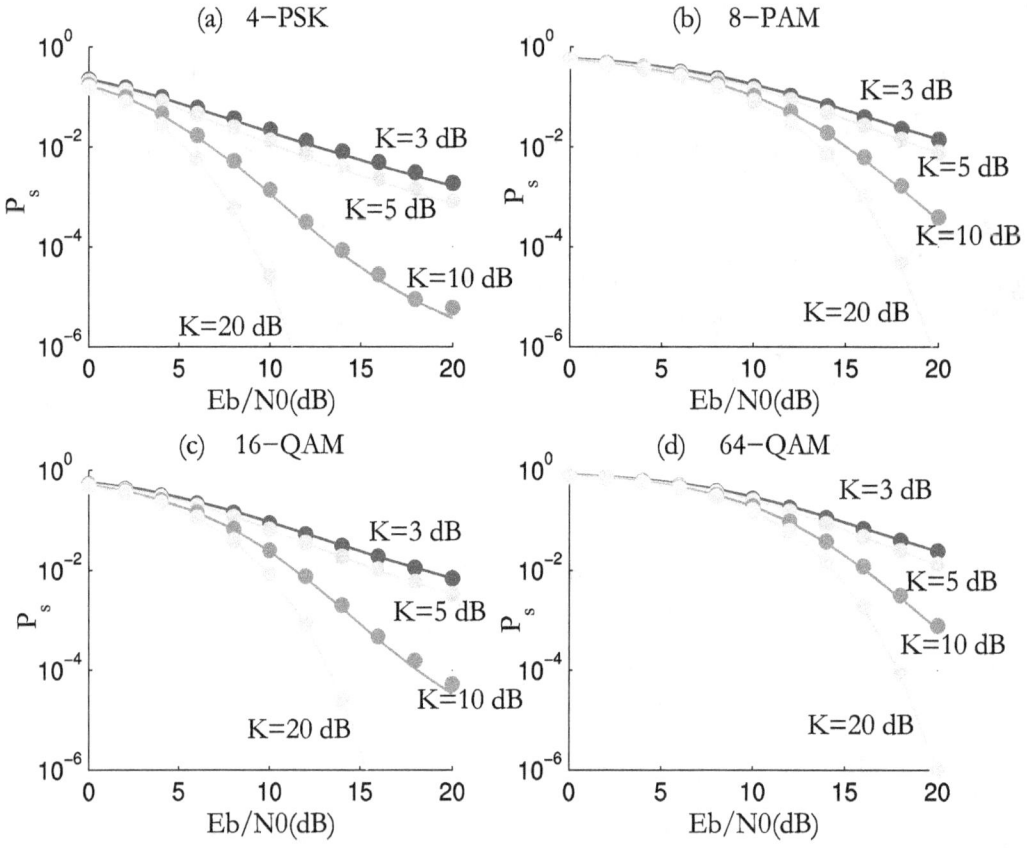

Fig. 4.9: Performance of modulations over Rician flat fading channel with different K factors : (a) QPSK , (b) 8-PAM, (c) 16-QAM, (d) 64-QAM

References

1. D. Tse and P. Viswanath, *Fundamentals of Wireless Communication*, Cambridge University Press, 2005
2. Andrea Goldsmith, *Wireless Communications*, Cambridge University Pres, first edition, August 8, 2005
3. M. K. Simon and M.-S. Alouini, *Digital Communication over Fading Channels A Unified Approach to Performance Analysis*, Wiley 2000
4. C. Tepedelenlioglu, A. Abdi, and G. B. Giannakis, *The Ricean K factor: Estimation and performance analysis*, IEEE Trans. Wireless Communication ,vol. 2, no. 4, pp. 799–810, Jul. 2003.
5. R. F. Lopes, I. Glover, M. P. Sousa, W. T. A. Lopes, and M. S. de Alencar, *A software simulation framework for spectrum sensing*, 13th International Symposium on Wireless Personal Multimedia Communications (WPMC 2010), Out. 2010.
6. M. C. Jeruchim, P. Balaban, and K. S. Shanmugan, *Simulation of Communication Systems, Methodology, Modeling, and Techniques*, second edition Kluwer Academic Publishers, 2000.

Chapter 5
Linear Equalizers

Abstract In communication systems that are dominated by inter-symbol interference, equalizers play an important role in combating unwanted distortion effects induced by the channel. Linear equalizers are based on linear time-invariant filters that are simple to analyze using conventional signal processing theory. This chapter is focused on design and implementation of two important types of linear equalizers namely, the *zero-forcing equalizer* and the *minimum mean squared error equalizer*.

5.1 Introduction

Transmission of signals through band-limited channels are vulnerable to the time dispersive effects of the channel that could manifest as *intersymbol interference* (ISI). Three major approaches to deal with ISI were introduced in the previous chapter and are reproduced here for reference.

- *Nyquist first criterion*: Force ISI effect to zero by signal design - pulse shaping techniques like sinc filtering, raised-cosine filtering and square-root raised-cosine filtering.
- *Partial response signaling*: Introduce controlled amount of ISI in the transmit side and prepare to deal with it at the receiver.
- *Design algorithms to counter ISI*: Learn to live with the presence of ISI and design robust algorithms at the receiver - Viterbi algorithm (maximum likelihood sequence estimation), equalizer etc.,

This chapter focuses on the third item, particularly on the technique of employing an *equalizer* at the receiver. An equalizer is a digital filter at the receiver that mitigates the distortion effects of *intersymbol interference* (ISI), introduced by a time-dispersive channel. If the channel is unknown and time varying, optimum design of transmit and receive filters is not possible. However, one can neutralize (equalize) the channel effects, if the impulse response of the channel is made known at the receiver. The problem of recovering the input signal in a time-dispersive channel boils down to finding the inverse of the channel and using it to neutralize the channel effects. However, finding the inverse of the channel impulse response may be difficult due to the following reasons:

- If the channel response is zero at any frequency, the inverse of the channel response is undefined at that frequency.
- Generally, the receiver does not know the channel response and extra circuitry (channel estimation) may be needed to estimate it.
- The channel can be varying in real time and therefore, the equalization needs to be adaptive.

Additionally, even if the channel response is of finite length, the equalizer can have infinite impulse response. Usually, from stability perspective, the equalizer response has to be truncated to a finite length. As a result, perfect equalization is difficult to achieve and does not always provide the best performance.

Classification of equalizer structures

Based on implementation structures, the equalizer structures are categorized as follows:

- Maximum Likelihood Sequence Estimator (MLSE)

  - Provides the most optimal performance.
  - Computational complexity can be extremely high for practical applications.

- Decision Feedback Equalizer

  - A nonlinear equalizer that employs previous decisions as training sequences.
  - Provides better performance but suffers from the serious drawback of error propagation in the event of erroneous decisions.
  - To provide robustness, it is often combined with training sequence based feed-forward equalization.

- Linear Equalizer

  - The most suboptimum equalizer with lower complexity.

5.2 Linear equalizers

The MLSE equalizers provide the most optimum performance among all the equalizer categories. For long channel responses, the MLSE equalizers become too complex and hence, from practical implementation point of view, other suboptimum filters are preferred. The most suboptimum equalizer is the *linear equalizer* that offers lower complexity. Linear equalizers are particularly effective for those channels where the ISI is not severe. Commonly, equalizers are implemented as a digital filter structure having a number of taps with complex tap coefficients (also referred as tap weights).

Choice of adaptation of equalizer tap weights

Based on the adaptation of tap coefficients, the linear equalizers can be classified into two types.

- *Preset equalizers*: If the channel is considered to be time-invariant during the data transmission interval, the equalizer weights are calculated during the beginning of the transmission session and are fixed during the rest of the transmission session.
- *Adaptive equalizers*: For time variant channels, the coefficients are computed using an adaptive algorithm to adapt for the change in the channel conditions, thus rendering the name. There exist numerous choices of algorithms for calculating and updating the adaptive weights - *least mean squares*, *recursive lattice square*, *conventional Kalman*, *square-root Kalman* and *fast Kalman* - to name a few.

Choice of sampling time

The choice of sampling time with respect to the symbol time period, affects the design and performance of equalizers. As a consequence, two types of equalization techniques exist.

- *Symbol Spaced Linear Equalizer*: The downsampler before the equalizer operates at 1-sample per symbol time (T_{sym}), as shown in Figure 5.1. These equalizers are very sensitive to sampling time, potentially introducing destructive aliasing effects and therefore the performance may not be optimal.

- *Fractionally Spaced Linear Equalizer*: Here, the idea is to reduce the space between the adjacent equalizer taps to a fraction of the symbol interval. Equivalently, the sampler that precedes the equalizer operates at a rate higher than symbol rate (say spacing in time kT_{sym}/M - for some integers k and M). Finally, the equalizer output is downsampled by symbol time, before delivering the output symbols to the detector. This is illustrated in Figure 5.2. Though it requires more computation, fractionally-spaced equalizers simplify the rest of the demodulator design [1] and are able to compensate for any arbitrary sampling time phase [2].

This chapter deals only with symbol-spaced linear equalizer with preset tap weights and adaptive weights using least mean squares algorithm. The treatment of fractionally-spaced equalizers is beyond the scope of this book.

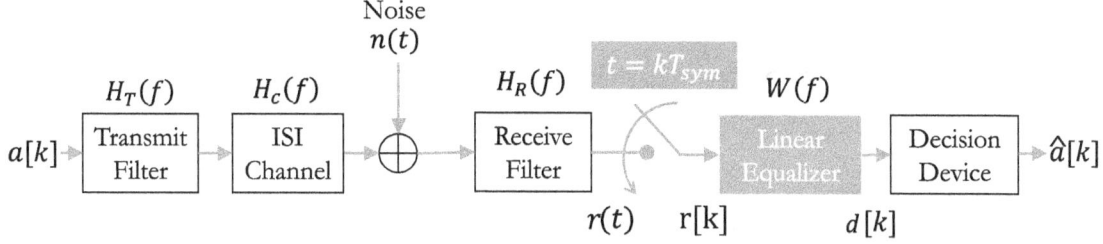

Fig. 5.1: Continuous-time channel model with symbol-spaced linear equalizer

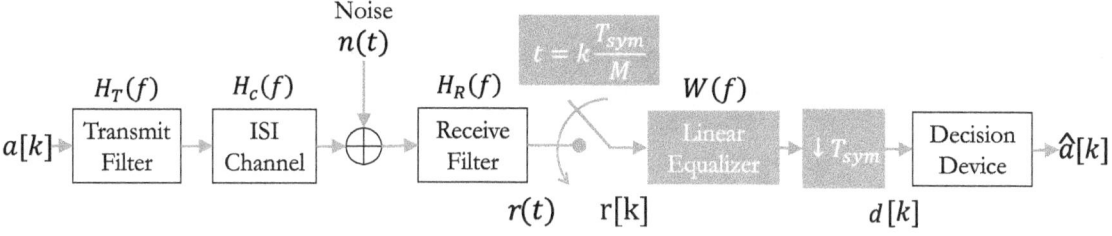

Fig. 5.2: Continuous-time channel model with fractionally spaced linear equalizer

Choice of criterion for tap weights optimization

The equalizer tap weights are usually chosen based on some optimal criterion of which two of them are given.

- *Peak distortion criterion*: Aims at minimizing the maximum possible distortion of a signal, caused by ISI, at the output of the equalizer. This is also equivalent to *zero forcing criterion*. This forms the basis for *zero forcing equalizers* [3].
- *Mean square error criterion*: Attempts to minimize the mean square value of an error term computed from the difference between the equalizer output and the transmitted information symbol. This forms the basis for *linear minimum mean square error (LMMSE)* equalizer and *least mean square (LMS)* algorithm.

5.3 Symbol spaced linear equalizer channel model

Consider the continuous-time channel model in Figure 5.1. In this generic representation, the transmit pulse shaping filter and the receiver filter are represented by the filters $H_T(f)$ and $H_R(f)$ respectively. The transmit filter may or may not be a square-root raised cosine filter. However, we assume that the receive filter is a square-root raised cosine filter. Therefore, the sampled noise $n[k] = h_R(t) * n(t) |_{t=kT_{sym}}$ is white Gaussian.

In general, the magnitude response of the channel $|H_c(f)|$ is not a constant over a range of frequencies channel, that is the impulse response of the channel $h_c(t)$ is not ideal. Therefore, linear distortions become unavoidable.

We designate the combined effect of transmit filter, channel and the receive filter as overall channel. The overall impulse response $h(t)$ is given by

$$h(t) = h_T(t) * h_c(t) * h_R(t) \tag{5.1}$$

The received signal after the symbol rate sampler is given by

$$r[k] = \sum_{i=-\infty}^{\infty} h[i]a[k-i] + z[k] \tag{5.2}$$

The resulting simplified discrete-time channel model is given in Figure 5.3, where $a[k]$ represent the information symbols transmitted through a channel having an arbitrary impulse response $h[k]$, $r[k]$ are the received symbols at the receiver and $n[k]$ is white Gaussian noise.

$$r[k] = \sum_{i=-\infty}^{\infty} h[i]a[k-i] + n[k] \tag{5.3}$$

Usually, in practice, the channel impulse response can be truncated to some finite length L. Under assumptions of causality of transmit filter, channel and the receive filter, we can safely hold that $h[i] = 0$ for $i < 0$. If L is sufficiently large, $h[i] \approx 0$ holds for $i \geq L$. Hence the received signal can be rewritten as

$$r[k] = \sum_{i=0}^{L-1} h[i]a[k-i] + n[k] \tag{5.4}$$

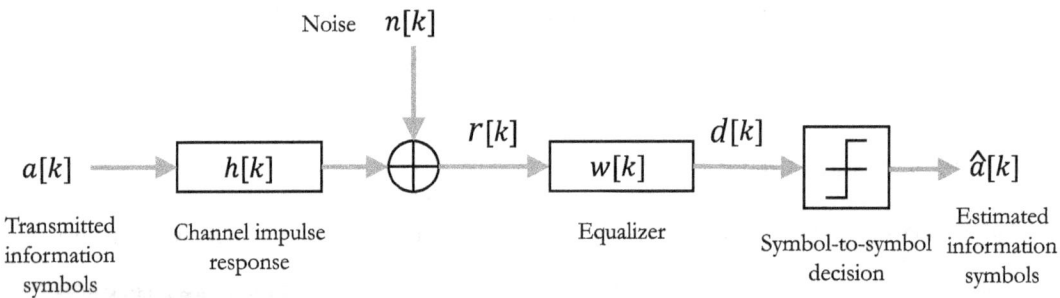

Fig. 5.3: Discrete-time channel model for symbol-spaced linear equalization

The equalizer co-efficients are represented by $w[k]$. Symbol-by-symbol decisions are made on the equalizer output $d[k]$ and the transmitted information symbols are estimated as $\hat{a}[k]$.

Figure 5.4 shows the equivalent frequency domain representation the channel model, where the frequency response can be computed using Z-transform as

$$W(z) = \sum_{k=-\infty}^{\infty} w[k]z^{-k} \tag{5.5}$$

A linear equalizer is the most simplest type of equalizer, that attempts to invert the channel transfer function $H(z)$ and filters the received symbols with the inverted response. Subsequently, using a symbol-by-symbol decision device, the filtered equalizer outputs are used to estimate the information symbols.

The structure of a linear equalizer can be of FIR or IIR type. The equalizer tap weights are calculated based an optimization criterion. We will be focusing on *zero-forcing (ZF)* criterion and the *minimum mean square error (MMSE)* criterion.

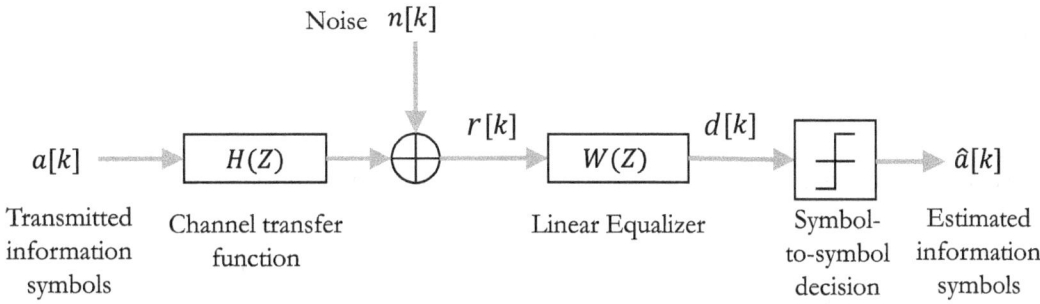

Fig. 5.4: Equivalent channel model in frequency domain with linear equalizer

5.4 Zero-forcing equalizer

A zero-forcing (ZF) equalizer is so named because it forces the residual ISI in the equalizer output $d[k]$ to zero. An optimum zero-forcing equalizer achieves perfect equalization by forcing the residual ISI at sampling instants kT except $k = 0$. This goal can be achieved if we allow for equalizer with infinite impulse response (IIR) structure. In most practical applications, the channel transfer function $H(z)$ can be approximated to finite response (FIR) filter and hence its perfect equalizing counterpart $W(z)$ will be an IIR filter.

$$W(z) = \frac{1}{H(z)} \tag{5.6}$$

As a result, the resulting overall transfer function, denoted by $Q(z)$, is given by

$$Q(z) = H(z)W(z) = 1 \tag{5.7}$$

Otherwise, in time-domain, the zero-forcing solution forces the overall response to zero at all positions except for the position at k_0 where its value is equal to Dirac delta function.

$$q[k] = h[k] * w[k] = \delta[k - k_0] = \begin{cases} 1, & for\ k = k_0 \\ 0, & for\ k \neq k_0 \end{cases} \tag{5.8}$$

For practical implementation, due to their inherent stability and better immunization against finite length word effects, FIR filters are practically preferred over IIR filters (refer chapter chapter 1 section 1.9 for more details). Implementing a zero-forcing equalizer as FIR filter would mean imposing causality and a length constraint on the equalizer filter whose transfer function takes the form

$$W(z) = \sum_{k=0}^{N-1} w[k]z^{-k} \tag{5.9}$$

Given the channel impulse response of length L and the equalizer filter of length N, the zero-forcing solution in equation 5.8, can be expressed as follows (refer chapter 1 section 1.6).

$$q[k] = h[k] * w[k] = \sum_{i=-\infty}^{\infty} h[i]w[k-i] = \delta[k-k_0], \qquad k = 0, \cdots, L+N-2 \tag{5.10}$$

In simple terms, with the zero-forcing solution, the overall impulse response takes the following form.

$$q[k] = h[k] * w[k] = \delta[k-k_0] = [0, 0, \cdots, 0, 1, 0, \cdots, 0, 0] \tag{5.11}$$

where, $q[k] = 1$ at position k_0.

The convolution sum of finite length sequence $\mathbf{h}$ of length L and a causal finite length sequence $\mathbf{w}$ of length N can be expressed as in matrix form as

$$\begin{bmatrix} q[0] \\ q[1] \\ q[2] \\ \vdots \\ q[L+N-2] \end{bmatrix} = \begin{bmatrix} h[0] & h[-1] & \cdots & h[-(N-1)] \\ h[1] & h[0] & \cdots & h[-(N-2)] \\ h[2] & h[1] & \cdots & h[-(N-3)] \\ \vdots & \vdots & \ddots & \vdots \\ h[L+N-2] & h[L+N-3] & \cdots & h[L-1] \end{bmatrix} \cdot \begin{bmatrix} w[0] \\ w[1] \\ w[2] \\ \vdots \\ w[N-1] \end{bmatrix} \tag{5.12}$$

Obviously, the zero-forcing solution in equation 5.8 can be expressed in matrix form as

$$\begin{bmatrix} h[0] & h[-1] & \cdots & h[-(N-1)] \\ h[1] & h[0] & \cdots & h[-(N-2)] \\ h[2] & h[1] & \cdots & h[-(N-3)] \\ \vdots & \vdots & \ddots & \vdots \\ h[L+N-2] & h[L+N-3] & \cdots & h[L-1] \end{bmatrix} \cdot \begin{bmatrix} w[0] \\ w[1] \\ w[2] \\ \vdots \\ w[N-1] \end{bmatrix} = \begin{bmatrix} 0 \\ \vdots \\ 0 \\ 1 \\ 0 \\ \vdots \\ 0 \end{bmatrix}$$

$$\mathbf{H}.\mathbf{w} = \delta_{k_0} \tag{5.13}$$

If we assume causality of the channel impulse response, $h[k] = 0$ for $k < 0$ and if L is chosen sufficiently large $h[k] \approx 0$ holds for $k \geq L$. Then, the zero-forcing solution can be simplified as

$$\begin{bmatrix} h[0] & 0 & \cdots & 0 & 0 \\ h[1] & h[0] & \cdots & 0 & 0 \\ h[2] & h[1] & \cdots & 0 & 0 \\ \vdots & \vdots & \ddots & \vdots & \vdots \\ 0 & 0 & \ddots & h[L-1] & 0 \\ 0 & 0 & \cdots & 0 & h[L-1] \end{bmatrix} \cdot \begin{bmatrix} w[0] \\ w[1] \\ w[2] \\ \vdots \\ w[N-2] \\ w[N-1] \end{bmatrix} = \begin{bmatrix} 0 \\ \vdots \\ 0 \\ 1 \\ 0 \\ \vdots \\ 0 \end{bmatrix}$$

$$\mathbf{T}(h).\mathbf{w} = \delta_{k_0}$$
$$\mathbf{H}.\mathbf{w} = \delta_{k_0} \tag{5.14}$$

where, $\mathbf{H} = \mathbf{T}(h)$ takes the form of a *Toeplitz matrix* (see chapter 1 section 1.7.2).

If the zero-forcing equalizer is assumed to perfectly compensate the channel impulse response, the output of the equalizer will be a Dirac delta function that is delayed by certain symbols due to equalizer delay (k_0). In matrix form, this condition is represented as

$$\mathbf{H} \cdot \mathbf{w} = \delta_{k_0} \tag{5.15}$$

where $\mathbf{H}$ is a rectangular *channel matrix* as defined in equations 5.13 and 5.14, $\mathbf{w}$ is a column vector containing the equalizer tap weights and δ_{k_0} is a column vector representing the Dirac delta function, with all elements equal to zero except for the unity element at $(k_0)^{th}$ position. The position of the unity element in the δ_{k_0} matrix determines the equalizer delay (k_0). The equalizer delay allows for end-to-end system delay. Without the equalizer delay, it would be difficult to compensate the effective channel delay using a causal FIR equalizer.

5.4.1 Least squares solution

The method of *least squares* can be applied to solve an overdetermined system of linear equations of form $\mathbf{Hw} = \delta_{k_0}$, that is, a system in which the $\mathbf{H}$ matrix is a rectangular $(L+N-1) \times N$ matrix with more equations than unknowns $((L+N-1) > N)$.

In general, for the overdetermined $(L+N-1) \times N$ system $\mathbf{Hw} = \delta_{k_0}$, there are solutions $\mathbf{w}$ minimizing the *Euclidean distance* $\left\| \mathbf{Hw} - \delta_{k_0} \right\|^2$ and those solutions are given by the square $N \times N$ system

$$\mathbf{H}^H \mathbf{Hw} = \mathbf{H}^H \delta_{k_0} \tag{5.16}$$

where $\mathbf{H}^H$ represents the *Hermitian transpose* (conjugate transpose of a matrix) of the channel matrix $\mathbf{H}$.

Besides, when the columns of $\mathbf{H}$ are linearly independent, it is apparent that $\mathbf{H}^H\mathbf{H}$ is symmetric and invertible, and so the solution for the zero-forcing equalizer coefficients $\mathbf{w}$ is unique and given by

$$\mathbf{w} = \left(\mathbf{H}^H\mathbf{H} \right)^{-1} \mathbf{H}^H \delta_{k_0} = \mathbf{H}^\dagger \delta_{k_0} \tag{5.17}$$

The expression $\mathbf{H}^\dagger = \left(\mathbf{H}^H\mathbf{H} \right)^{-1} \mathbf{H}^H$ is called *pseudo-inverse matrix* or *Moore-Penrose generalized matrix inverse*. To solve for equalizer taps using equation 5.17, the channel matrix $\mathbf{H}$ has to be estimated.

The *mean square error (MSE)* between the equalized samples and the ideal samples is calculated as [5]

$$\xi_N = 1 - \sum_{i=0}^{k_0} w_N(i) h_c^*(k_0 - i) \tag{5.18}$$

The equations for zero forcing equalizer have the same form as that of MMSE equalizer (see equations 5.39 and 5.40) , except for the noise term. Therefore, the MSE of the ZF equalizer can be written as [6] [7]

$$\xi_{min} = \sigma_a^2 \left(1 - \delta_{k_0}^T \mathbf{H} \mathbf{H}^\dagger \delta_{k_0}\right) \tag{5.19}$$

The optimal-delay that minimizes the MSE is simply the index of the maximum diagonal element of the matrix $\mathbf{H} \mathbf{H}^\dagger$

$$optimum\ delay = k_{opt} = \arg \max_{index} \left[diag \left(\mathbf{H} \mathbf{H}^\dagger \right) \right] \tag{5.20}$$

5.4.2 Noise enhancement

From the given discrete-time channel model (Figure 5.3), the received sequence $\mathbf{r}$ can be computed as the convolution sum of the input sequence $\mathbf{a}$ of length N and the channel impulse response $\mathbf{h}$ of length L

$$r[k] = h[k] * a[k] + n[k] = \sum_{i=-\infty}^{\infty} h[i]a[k-i] + n[k] \qquad , k = 0, 1, \cdots, L+N-2 \tag{5.21}$$

Following the derivations similar to equations 5.10 and 5.12, the convolution sum can be expressed in matrix form as

$$\begin{bmatrix} r[0] \\ r[1] \\ r[2] \\ \vdots \\ r[L+N-2] \end{bmatrix} = \begin{bmatrix} h[0] & h[-1] & \cdots & h[-(N-1)] \\ h[1] & h[0] & \cdots & h[-(N-2)] \\ h[2] & h[1] & \cdots & h[-(N-3)] \\ \vdots & \vdots & \ddots & \vdots \\ h[L+N-2] & h[L+N-3] & \cdots & h[L-1] \end{bmatrix} \cdot \begin{bmatrix} a[0] \\ a[1] \\ a[2] \\ \vdots \\ a[N-1] \end{bmatrix} + \begin{bmatrix} n[0] \\ n[1] \\ n[2] \\ \vdots \\ n[N-1] \end{bmatrix} \tag{5.22}$$

$$\mathbf{r} = \mathbf{H}.\mathbf{a} + \mathbf{n} \tag{5.23}$$

If a linear zero-forcing equalizer is used, obtaining the equalizer output is equivalent to applying the pseudo-inverse matrix $\mathbf{H}^\dagger$ on the vector of received signal sequence $\mathbf{r}$.

$$\mathbf{H}^\dagger \mathbf{r} = \mathbf{H}^\dagger \mathbf{H}.\mathbf{a} + \mathbf{H}^\dagger \mathbf{n} = \mathbf{a} + \tilde{\mathbf{n}} \tag{5.24}$$

This renders the additive white Gaussian noise become colored and thereby complicates optimal detection.

5.4.3 Design and simulation of zero forcing equalizer

The simulation model for ZF equalizer is shown in Figure 5.3. The simulation model involves the overall continuous time channel impulse response $h(t)$ with discrete-time equivalent $h[k]$ and a discrete-time zero forcing symbol spaced equalizer $w[k]$. The goal is to design the zero forcing equalizer of length N that could compensate for the distortion effects introduced by the channel of length L.

As an example for simulation, the sampling frequency of the overall system is assumed as $F_s = 100Hz$ with 5 samples per symbol ($nSamp = 5$) and a channel model with the following overall channel impulse response at baud-rate is also assumed.

$$h(t) = \frac{1}{1 + \left(\frac{t}{T_{sym}}\right)^2} \tag{5.25}$$

Program 5.1: Channel Model

```
1   %System parameters
2   nSamp=5; %Number of samples per symbol determines baud rate Tsym
3   Fs=100; %Sampling Frequency of the system
4   Ts=1/Fs; %Sampling time
5   Tsym=nSamp*Ts; %symbol time period
6
7   %Define transfer function of the channel
8   k=6; %define limits for computing channel response
9   N0 = 0.001; %Standard deviation of AWGN channel noise
10  t = -k*Tsym:Ts:k*Tsym; %time base defined till +/-kTsym
11  h_t = 1./(1+(t/Tsym).^2);%channel model, replace with your own model
12  h_t = h_t + N0*randn(1,length(h_t));%add Noise to the channel response
13  h_k = h_t(1:nSamp:end);%downsampling to represent symbol rate sampler
14  t_inst=t(1:nSamp:end); %symbol sampling instants
```

Next, plot the channel impulse response of the above model, the resulting plot is given in Figure 5.5.

Program 5.2: Plot channel Impulse Response

```
1   figure; plot(t,h_t); hold on; %channel response at all sampling instants
2   stem(t_inst,h_k,'r'); %channel response at symbol sampling instants
3   legend('continuous-time model','discrete-time model');
4   title('Channel impulse response'); xlabel('Time (s)');ylabel('Amplitude');
```

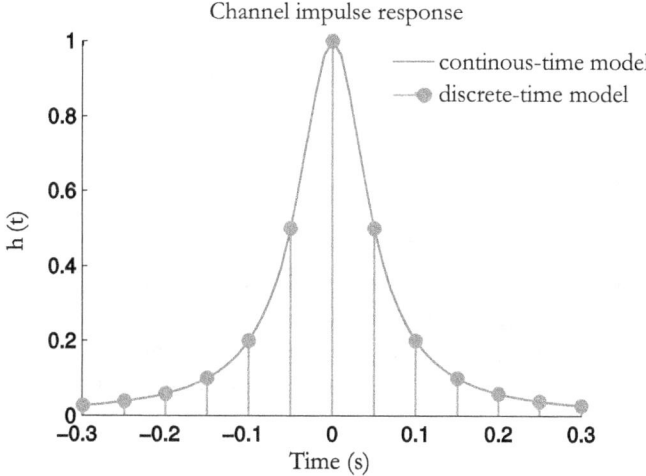

Fig. 5.5: Channel Impulse Response

Since the channel response, plotted in Figure 5.5, is of length $L = 13$, a zero forcing forcing filter of length $N > L$ is desired. Let's fix the length of the zero forcing filter to $N = 14$. The zero forcing equalizer design, follows equation 5.17 and the equalizer delay determines the position of the solitary ' 1 ' in the δ_{k_0} matrix that represents the Dirac-delta function.

Given the channel impulse response and the desired equalizer length, the following function designs a zero forcing equalizer using equations 5.17, 5.19 and 5.20.

Program 5.3: *zf_equalizer.m*: Function to design ZF equalizer and to equalize a signal

```
1   function [w,err,optDelay]=zf_equalizer(h,N,delay)
2   % Delay optimized zero forcing equalizer.
3   %  [w,err,optDelay]=zf_equalizer(h,N,delay) designs a ZERO FORCING equalizer
4   %  w for given channel impulse response h, the desired length of the equalizer
5   %  N and equalizer delay (delay). Also returns the equalizer error (err),the
6   %  best optimal delay (optDelay) that could work best for the designed equalizer
7   %
8   %  [w,err,optDelay]=zf_equalizer(h,N) designs a DELAY OPTIMIZED ZERO FORCING
9   %  equalizer w for given channel impulse response h, the desired length of the
10  %  equalizer N. Also returns equalizer error(err),the best optimal delay(optDelay)
11
12      h=h(:); %Channel matrix to solve simultaneous equations
13      L=length(h); %length of CIR
14      H=convMatrix(h,N); %(L+N-1)xN matrix - see Chapter 1
15      %section title - Methods to compute convolution
16
17      %compute optimum delay based on MSE
18      Hp = inv(H'*H)*H'; %Moore-Penrose Pseudo inverse
19      [~,optDelay] = max(diag(H*Hp));
20      optDelay=optDelay-1;%since Matlab index starts from 1
21      k0=optDelay;
22
23      if nargin--3,
24          if delay >=(L+N-1), error('Too large delay'); end
25          k0=delay;
26      end
27      d=zeros(N+L-1,1);
28      d(k0+1)=1; %optimized position of equalizer delay
29      w=Hp*d;%Least Squares solution
30      err=1-H(k0+1,:)*w; %equalizer error (MSE)-reference [5]
31      MSE=(1-d'*H*Hp*d);%MSE and err are equivalent
32  end
```

Let us design a zero forcing equalizer with $N = 14$ taps for the given channel response in Figure 5.5 and equalizer delay k_0 set to 11. The designed equalizer is tested by sending the channel impulse response as the input, $r[k] = h[k]$. The equalizer is expected to compensate for it completely as indicated in Figure 5.6. The overall system response is computed as the convolution of channel impulse response and the response of the zero forcing equalizer.

Program 5.4: Design and test the zero forcing equalizer

```
1   %Equalizer Design Parameters
2   N = 14; %Desired number of taps for equalizer filter
3   delay = 11;
4
5   %design zero-forcing equalizer for given channel and get tap weights
6   %and filter the input through the equalizer find equalizer co-effs for given CIR
7   [w,error,k0]=zf_equalizer(h_k,N,delay);
8   %[w,error,k0]=zf_equalizer(h,N,); %Try this delay optimized equalizer
9
```

```
10  r_k=h_k; %Test the equalizer with the sampled channel response as input
11  d_k=conv(w,r_k); %filter input through the eq
12  h_sys=conv(w,h_k); %overall effect of channel and equalizer
13
14  disp(['ZF Equalizer Design: N=', num2str(N), ...
15      ' Delay=',num2str(delay), ' Error=', num2str(error)]);
16  disp('ZF equalizer weights:'); disp(w);
```

Compute and plot the frequency response of the channel, equalizer and the overall system. The resulting plot, given in Figure 5.6, shows how the zero forcing equalizer has compensated the channel response by inverting it.

Program 5.5: Frequency responses of channel equalizer and overall system

```
1  [H_F,Omega_1]=freqz(h_k);  %frequency response of channel
2  [W,Omega_2]=freqz(w); %frequency response of equalizer
3  [H_sys,Omega_3]=freqz(h_sys); %frequency response of overall system
4
5  figure;
6  plot(Omega_1/pi,20*log(abs(H_F)/max(abs(H_F))),'g'); hold on;
7  plot(Omega_2/pi,20*log(abs(W)/max(abs(W))),'r');
8  plot(Omega_3/pi,20*log(abs(H_sys)/max(abs(H_sys))),'k');
9  legend('channel','ZF equalizer','overall system');
10  title('Frequency response');ylabel('Magnitude(dB)');
11  xlabel('Normalized frequency(x \pi rad/sample)');
```

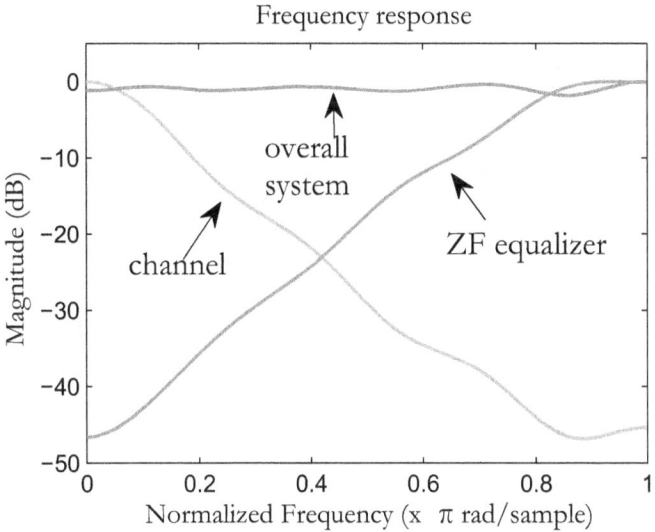

Fig. 5.6: Zero forcing equalizer compensates for the channel response

The response of the filter in time domain is plotted in Figure 5.7. The plot indicates that the equalizer has perfectly compensated the input when $r[k] = h[k]$.

Program 5.6: Design and test the zero forcing equalizer

```
1  figure; %Plot equalizer input and output
2  subplot(2,1,1); stem(0:1:length(r_k)-1,r_k);
3  title('Equalizer input'); xlabel('Samples'); ylabel('Amplitude');
4
5  subplot(2,1,2); stem(0:1:length(d_k)-1,d_k);
6  title(['Equalizer output- N=', num2str(N), ...
7      ' Delay=',num2str(delay), ' Error=', num2str(error)]);
8  xlabel('Samples'); ylabel('Amplitude');
```

The calculated zero forcing design error depends on the equalizer delay. For an equalizer designed with 14 taps and a delay of 11, the MSE error $=7.04 \times 10^{-4}$ for a noiseless input. On the other hand the error$=0.9985$ for delay$=1$. Thus the equalizer delay plays a vital role in the zero forcing equalizer design. Knowledge of the delay introduced by the equalizer is very essential to determine the position of the first valid sample at the output of the equalizer (see sections 5.6 and 5.7). The complete simulation code, that includes all the above discussed code snippets, is given next.

Program 5.7: *zf_equalizer_test.m*: Simulation of Zero Forcing equalizer

```
1  clearvars; clc;
2  %System parameters
3  nSamp=5; %Number of samples per symbol determines baud rate Tsym
4  Fs=100; %Sampling Frequency of the system
5  Ts=1/Fs; %Sampling time
6  Tsym=nSamp*Ts; %symbol time period
7
8  %Define transfer function of the channel
9  k=6; %define limits for computing channel response
10 N0 = 0.001; %Standard deviation of AWGN channel noise
11 t = -k*Tsym:Ts:k*Tsym; %time base defined till +/-kTsym
12 h_t = 1./(1+(t/Tsym).^2);%channel model, replace with your own model
13 h_t = h_t + N0*randn(1,length(h_t));%add Noise to the channel response
14 h_k = h_t(1:nSamp:end);%downsampling to represent symbol rate sampler
15 t_inst=t(1:nSamp:end); %symbol sampling instants
16
17 figure; plot(t,h_t); hold on; %channel response at all sampling instants
18 stem(t_inst,h_k,'r'); %channel response at symbol sampling instants
19 legend('continuous-time model','discrete-time model');
20 title('Channel impulse response'); xlabel('Time (s)');ylabel('Amplitude');
21
22 %Equalizer Design Parameters
23 N = 14; %Desired number of taps for equalizer filter
24 delay = 11;
25
26 %design zero-forcing equalizer for given channel and get tap weights
27 %and filter the input through the equalizer find equalizer co-effs for given CIR
28 [w,error,k0]=zf_equalizer(h_k,N,delay);
29 %[w,error,k0]=zf_equalizer(h,N,); %Try this delay optimized equalizer
30 r_k=h_k; %Test the equalizer with the sampled channel response as input
31 d_k=conv(w,r_k); %filter input through the eq
32 h_sys=conv(w,h_k); %overall effect of channel and equalizer
```

```
33
34   disp(['ZF equalizer design: N=', num2str(N), ...
35       ' Delay=',num2str(delay), ' error=', num2str(error)]);
36   disp('ZF equalizer weights:'); disp(w);
37
38   %Frequency response of channel,equalizer & overall system
39   [H_F,Omega_1]=freqz(h_k);  %frequency response of channel
40   [W,Omega_2]=freqz(w); %frequency response of equalizer
41   [H_sys,Omega_3]=freqz(h_sys); %frequency response of overall system
42
43   figure;
44   plot(Omega_1/pi,20*log(abs(H_F)/max(abs(H_F))),'g'); hold on;
45   plot(Omega_2/pi,20*log(abs(W)/max(abs(W))),'r');
46   plot(Omega_3/pi,20*log(abs(H_sys)/max(abs(H_sys))),'k');
47   legend('channel','ZF equalizer','overall system');
48   title('Frequency response');ylabel('Magnitude(dB)');
49   xlabel('Normalized frequency(x \pi rad/sample)');
50
51   figure; %Plot equalizer input and output(time-domain response)
52   subplot(2,1,1); stem(0:1:length(r_k)-1,r_k);
53   title('Equalizer input'); xlabel('Samples'); ylabel('Amplitude');
54   subplot(2,1,2); stem(0:1:length(d_k)-1,d_k);
55   title(['Equalizer output- N=', num2str(N), ...
56       ' Delay=',num2str(delay), ' Error=', num2str(error)]);
57   xlabel('Samples'); ylabel('Amplitude');
```

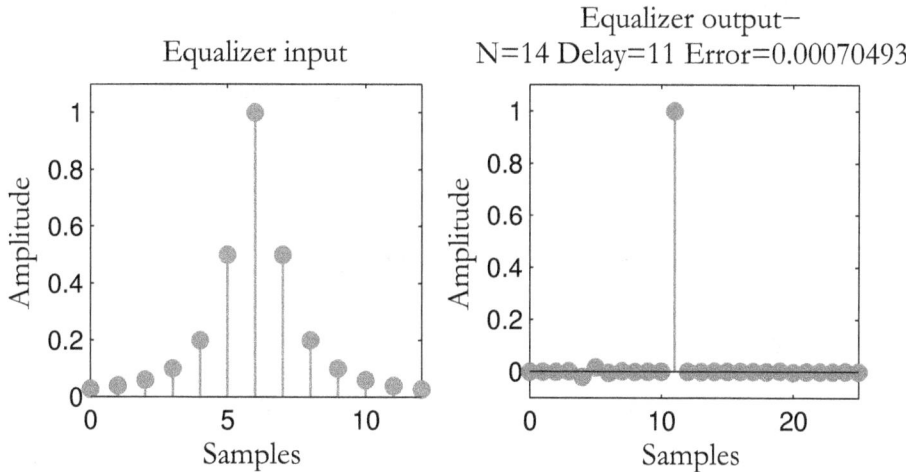

Fig. 5.7: Input samples and the output samples from designed zero-forcing equalizer

5.4.4 Drawbacks of zero forcing equalizer

Since the response of a zero forcing equalizer is the inverse of the channel response (refer Figure 5.6), the zero forcing equalizer applies large gain at those frequencies where the channel has severe attenuation or spectral nulls. As a result, the large gain also amplifies the additive noise at those frequencies and hence the zero forcing equalization suffers from *noise-enhancement*. Another issue is that the additive white noise (which can be added as part of channel model) can become colored and thereby complicating the optimal detection. The simulated results in Figure 5.8, (obtained by increasing standard deviation of additive noise), show the influence of additive channel noise on the equalizer output. These unwanted effects can be eliminated by applying *minimum mean squared-error* criterion.

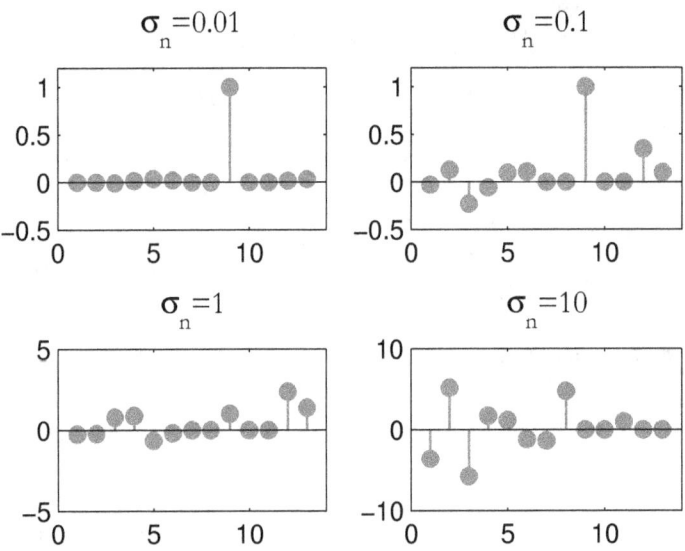

Fig. 5.8: Effect of additive noise on zero-forcing equalizer output

5.5 Minimum mean square error (MMSE) equalizer

The zero-forcing equalizer suffers from the ill-effects of noise enhancement. An improved performance over zero-forcing equalizer can be achieved by considering the noise term in the design of the equalizer. An MMSE filter design achieves this by accounting for a trade-off between noise enhancement and interference suppression.

Minimization of error variance and bias is the goal of the MMSE filter design problem, in other words, it tries to minimize the mean square error. In this design problem, we wish to design an FIR MMSE equalizer of length N having the following filter transfer function.

$$W(z) = \sum_{k=0}^{N-1} w[k]z^{-k} \tag{5.26}$$

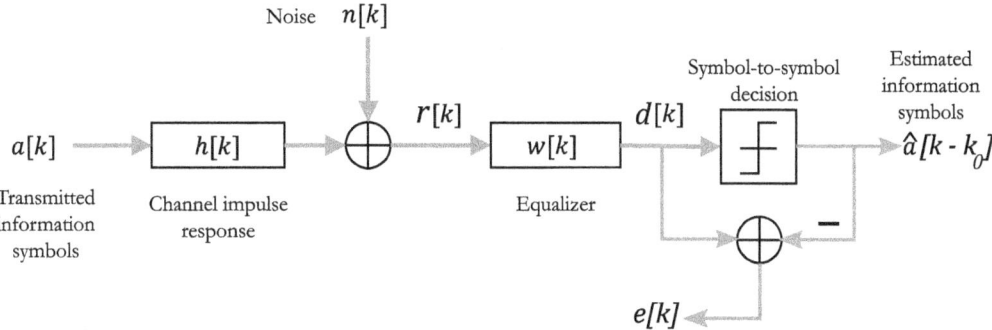

Fig. 5.9: Filter design model for MMSE symbol spaced linear equalizer

With reference to the generic discrete-time channel model given in Figure 5.9, an MMSE equalizer tries to minimize the variance and bias, from the following error signal

$$e[k] = d[k] - \hat{a}[k - k_0] \qquad (5.27)$$

To account for the delay due to the channel and equalizer, the discrete-time model (Figure 5.9) includes the delay k_0 at the decision device output. We note that the error signal depends on the estimated information symbols $\hat{a}[k - k_0]$. Since there is a possibility for erroneous decisions at the output of decision device, it is difficult to account for those errors. Hence, for filter optimization, in training phase, we assume $\hat{a}[k - k_0] = a[k - k_0]$.

$$e[k] = d[k] - a[k - k_0] \qquad (5.28)$$

The error signal can be rewritten to include the effect of an N length FIR equalizer on the received symbols

$$e[k] = d[k] - a[k - k_0] = \sum_{i=0}^{N-1} w[i]r[k-i] - a[k - k_0] \qquad (5.29)$$

Given the received sequence or noisy measurement $\mathbf{r}$, the goal is to design an equalizer filter $\mathbf{w}$ that would recover the information symbols $\mathbf{a}$. This is a classic *Weiner filter* problem. The Weiner filter design approach requires that we find the filter coefficients $\mathbf{w}$ that minimizes the mean square error ξ. This criterion can be concisely stated as

$$\mathbf{w} = arg \min \left(\xi \right) = arg \min E \left\{ \left| e[k] \right|^2 \right\} \qquad (5.30)$$

where $E[\cdot]$ is the expectation operator.

For complex signal case, the error signal given in equation 5.29 can be rewritten as

$$e[k] = d[k] - a[k - k_d] = \sum_{i=0}^{N-1} w^*[i]r[k-i] - a[k - k_0]$$

$$= \begin{bmatrix} w_0^* & w_1^* & \cdots & w_{N-1}^* \end{bmatrix} \begin{bmatrix} r_k \\ r_{k-1} \\ \vdots \\ r_{k-(N-1)} \end{bmatrix} - a[k - k_0]$$

$$= \mathbf{w}^H \mathbf{r} - a[k - k_0] \qquad (5.31)$$

where, $\mathbf{w} = [w_0, w_1, \cdots, w_{N-1}]^H$ are the complex filter coefficients with superscript H denoting Hermitian transpose and $\mathbf{r} = [r_k, r_{k-1}, \cdots, r_{k-N-1}]^T$ denotes the complex received sequence.

Using equation 5.31, the *mean square error (MSE)* can be derived as

$$
\begin{aligned}
\xi &= E\left\{|e[k]|^2\right\} = E\left\{e[k]e^*[k]\right\} \\
&= E\left\{\left(\mathbf{w}^H\mathbf{r} - a[k-k_0]\right)\left(\mathbf{w}^H\mathbf{r} - a[k-k_0]\right)^H\right\} \\
&= \mathbf{w}^H \underbrace{E\left\{\mathbf{rr}^H\right\}}_{\mathbf{R}_{rr}} \mathbf{w} - \mathbf{w}^H \underbrace{E\left\{\mathbf{r}a^*[k-k_0]\right\}}_{\mathbf{R}_{ra}} - E\left\{a[k-k_0]\mathbf{r}^H\right\}\mathbf{w} + E\left\{|a[k-k_0]|^2\right\} \\
&= \mathbf{w}^H\mathbf{R}_{rr}\mathbf{w} - \mathbf{w}^H\mathbf{R}_{ra} - \mathbf{R}_{ra}^H\mathbf{w} + E\left\{|a[k-k_0]|^2\right\}
\end{aligned}
\tag{5.32}
$$

with *autocorrelation matrix* $\mathbf{R}_{rr} = E\left\{\mathbf{r} \cdot \mathbf{r}^H\right\}$ and *crosscorrelation vector* $\mathbf{R}_{ra} = E\left\{\mathbf{r}a^*[k-k_0]\right\}$.

To find the complex tap weights $\mathbf{w}$ that provide the minimum MSE, the gradient of ξ is set to zero. The gradient of mean square error is given by

$$
\frac{\partial}{\partial\mathbf{w}^*}(\xi) = \frac{\partial}{\partial\mathbf{w}^*}\left(\mathbf{w}^H\mathbf{R}_{rr}\mathbf{w}\right) - \frac{\partial}{\partial\mathbf{w}^*}\left(\mathbf{w}^H\mathbf{R}_{ra}\right)
\tag{5.33}
$$

$$
- \frac{\partial}{\partial\mathbf{w}^*}\left(\mathbf{R}_{ra}^H\mathbf{w}\right) + \frac{\partial}{\partial\mathbf{w}^*}\left(E\left\{|a[k-k_0]|^2\right\}\right)
\tag{5.34}
$$

$$
= \mathbf{R}_{rr}\mathbf{w} - \mathbf{R}_{ra} - 0 + 0
\tag{5.35}
$$

Equating the gradient of MSE to zero, the Weiner solution to find the complex filter coefficients of the MMSE filter is given by

$$
\mathbf{w} = \mathbf{R}_{rr}^{-1}\mathbf{R}_{ra}
\tag{5.36}
$$

This equation is called as the *Weiner-Hopf equation* with the autocorrelation matrix $\mathbf{R}_{rr}$ and the cross correlation vector $\mathbf{R}_{ra}$ given by

$$
\mathbf{R}_{rr} = E\left\{\mathbf{rr}^H\right\} = \begin{bmatrix}
E\left\{r_k r_k^*\right\} & E\left\{r_k r_{k-1}^*\right\} & \cdots & E\left\{r_k r_{k-(N-1)}^*\right\} \\
E\left\{r_{k-1} r_k^*\right\} & E\left\{r_{k-1} r_{k-1}^*\right\} & \cdots & E\left\{r_{k-1} r_{k-(N-1)}^*\right\} \\
\vdots & \vdots & \ddots & \vdots \\
E\left\{r_{k-(N-1)} r_k^*\right\} & E\left\{r_{k-(N-1)} r_{k-1}^*\right\} & \cdots & E\left\{r_{k-(N-1)} r_{k-(N-1)}^*\right\}
\end{bmatrix}
\tag{5.37}
$$

$$
\mathbf{R}_{ra} = E\left\{\mathbf{r}a^*[k-k_0]\right\} = \left[E\left\{r_k a_{k-k_0}^*\right\} \ E\left\{r_{k-1} a_{k-k_0}^*\right\} \ \cdots \ E\left\{r_{k-(N-1)} a_{k-k_0}^*\right\}\right]^T
\tag{5.38}
$$

5.5.1 Alternate solution

An alternate form of equation to solve for MMSE equalizer taps [6] [7], that allow for end-to-end system delay is given by

$$
\mathbf{w} = \left(\mathbf{H}^H\mathbf{H} + \frac{\sigma_n^2}{\sigma_a^2}\mathbf{I}\right)^{-1}\mathbf{H}^H\delta_{k_0}
\tag{5.39}
$$

Here, $\mathbf{H}$ is the channel matrix as defined in equation 5.14, $\mathbf{H}^H$ its *Hermitian transpose*, $\mathbf{I}$ is an identity matrix, the ratio $\frac{\sigma_n^2}{\sigma_a^2}$ is the inverse of signal-to-noise ratio (SNR) and the column vector $\delta_{k_0} = [\cdots, 0, 1, 0, \cdots]^T$ with a 1 at k_0th position. This equation is very similar to the equation 5.17 of the zero-forcing equalizer, except for the signal-to-noise ratio term. Therefore, for a noiseless channel, the solution is identical for the MMSE equalizers and the zero forcing equalizers. This equation allows for end-to-end system delay. Without the equalizer delay (k_0), it is difficult to compensate for the effective channel delay using a causal FIR equalizer. The solution for the MMSE equalizer depends on the availability and accuracy of the channel matrix $\mathbf{H}$, the knowledge of the variances σ_a^2 and σ_n^2 (which is often the hardest challenge), and the invertibility of the matrix $\mathbf{HH}^H + \frac{\sigma_n^2}{\sigma_a^2}\mathbf{I}$. In this section, it is assumed that the receiver possesses the knowledge of the channel matrix $\mathbf{H}$ and the variances - σ_a^2, σ_n^2, and our goal is to design a delay-optimized MMSE equalizer.

The minimum MSE is given by [6] [7]

$$\xi_{min} = \sigma_a^2 \left(1 - \delta_{k_0}^T \mathbf{H} \left[\mathbf{H}^H \mathbf{H} + \frac{\sigma_n^2}{\sigma_a^2}\mathbf{I} \right]^{-1} \mathbf{H}^H \delta_{k_0} \right) \tag{5.40}$$

The optimal-delay that minimizes the MSE is simply the index of the maximum diagonal element of the matrix $\mathbf{H} \left[\mathbf{H}^H \mathbf{H} + \frac{\sigma_n^2}{\sigma_a^2}\mathbf{I} \right]^{-1} \mathbf{H}^H$.

$$optimum\ delay = k_{opt} = \underset{index}{\arg\max} \left[diag \left(\mathbf{H} \left[\mathbf{H}^H \mathbf{H} + \frac{\sigma_n^2}{\sigma_a^2}\mathbf{I} \right]^{-1} \mathbf{H}^H \right) \right] \tag{5.41}$$

5.5.2 Design and simulation of MMSE equalizer

A Matlab function for the design of delay-optimized MMSE equalizer is given next. This function is based on the equations 5.39, 5.40 and 5.41.

Program 5.8: *mmse_equalizer.m*: Function to design a delay-optimized MMSE equalizer

```
function [w,mse,optDelay]=mmse_equalizer(h,snr,N,delay)
% Delay optimized MMSE Equalizer.
%   [w,mse]=mmse_equalizer(h,snr,N,delay) designs a MMSE equalizer w
%   for given channel impulse response h, input 'snr' (dB), the desired length
%   of the equalizer N and equalizer delay (delay). It also returns the mean
%   square error (mse) and the optimal delay (optDelay) of the designed equalizer.
%
%   [w,mse,optDelay]=mmse_equalizer(h,snr,N) designs a DELAY OPTIMIZED MMSE
%   equalizer w for given channel impulse response h, input 'snr' (dB) and
%   the desired length of the equalizer N. Also returns the mean square error(mse)
%   and the optimal delay (optDelay) of the designed equalizer.

    h=h(:);%channel matrix to solve for simultaneous equations
    L=length(h); %length of CIR
    H=convMatrix(h,N); %(L+N-1)xN matrix - see Chapter 1
    %section title - Methods to compute convolution

```

```
18   gamma = 10^(-snr/10); %inverse of SNR
19   %compute optimum delay
20   [~,optDelay] = max(diag(H*inv(H'*H+gamma*eye(N))*H'));
21   optDelay=optDelay-1; %since Matlab index starts from 1
22   k0=optDelay;
23
24   if nargin==4,
25     if delay >=(L+N-1), error('Too large delay'); end
26     k0=delay;
27   end
28   d=zeros(N+L-1,1);
29   d(k0+1)=1; %optimized position of equalizer delay
30   w=inv(H'*H+gamma*eye(N))*H'*d; %Least Squares solution
31   mse=(1-d'*H*inv(H'*H+gamma*eye(N))*H'*d);%assume var(a)=1
32   end
```

The simulation methodology to test the zero forcing equalizer was discussed in the previous section. The same methodology is applied here to test MMSE equalizer. Here, the same channel model with ISI length $L = 13$ is used with slightly increased channel noise ($\sigma_n^2 = 0.1$). The channel response at symbol sampling instants are shown in Figure 5.10.

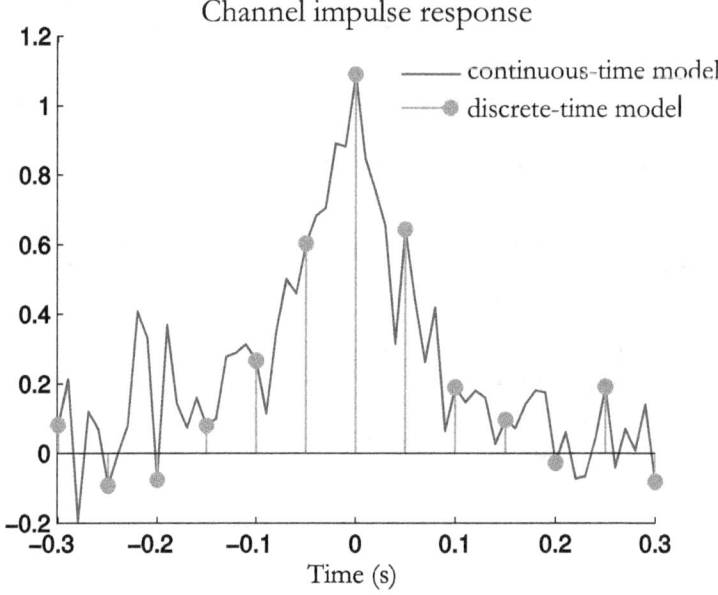

Fig. 5.10: A noisy channel with CIR of length $N = 14$

Next, a delay-optimized MMSE equalizer of length $N = 14$ is designed and used for compensating the distortion introduced by the channel. Figure 5.11 shows the MMSE equalizer in action that beautifully compensates the noisy distorted data. The complete simulation code is given next.

Program 5.9: *mmse_equalizer_test.m*: Simulation of MMSE equalizer

```
1   clearvars clc;
2   %System parameters
3   nSamp=5; %Number of samples per symbol determines baud rate Tsym
4   Fs=100; %Sampling Frequency of the system
5   Ts=1/Fs; %Sampling period
6   Tsym=nSamp*Ts;
7
8   %Define transfer function of the channel
9   k=6; %define limits for computing channel response
10  N0 = 0.1; %Standard deviation of AWGN noise
11  t = -k*Tsym:Ts:k*Tsym; %time base defined till +/-kTsym
12  h_t = 1./(1+(t/Tsym).^2); %channel model, replace with your own model
13  h_t = h_t + N0*randn(1,length(h_t)); %add Noise to the channel response
14  h_k= h_t(1:nSamp:end);%downsampling to represent symbol rate sampler
15  t_k=t(1:nSamp:end); %symbol sampling instants
16
17  figure;plot(t,h_t);hold on;%channel response at all sampling instants
18  stem(t_k,h_k,'r'); %channel response at symbol sampling instants
19  legend('continuous-time model','discrete-time model');
20  title('Channel impulse response');
21  xlabel('Time (s)');ylabel('Amplitude');
22
23  %Equalizer Design Parameters
24  nTaps = 14; %Desired number of taps for equalizer filter
25
26  %design DELAY OPTIMIZED MMSE eq. for given channel, get tap weights
27  %and filter the input through the equalizer
28  noiseVariance = N0^2; %noise variance
29  snr = 10*log10(1/N0); %convert to SNR (assume var(signal) = 1)
30  [w,mse,optDelay]=mmse_equalizer(h_k,snr,nTaps); %find eq. co-effs
31
32  r_k=h_k; %Test the equalizer with the channel response as input
33  d_k=conv(w,r_k); %filter input through the equalizer
34  h_sys=conv(w,h_k); %overall effect of channel and equalizer
35
36  disp(['MMSE equalizer design: N=', num2str(nTaps),' delay=',num2str(optDelay)])
37  disp('MMSE equalizer weights:'); disp(w)
38
39  figure; subplot(1,2,1); stem(0:1:length(r_k)-1,r_k);
40  xlabel('Samples'); ylabel('Amplitude');title('Equalizer input');
41  subplot(1,2,2); stem(0:1:length(d_k)-1,d_k);
42  xlabel('Samples'); ylabel('Amplitude');
43  title(['Equalizer output- N=', num2str(nTaps),' delay=',num2str(optDelay)]);
```

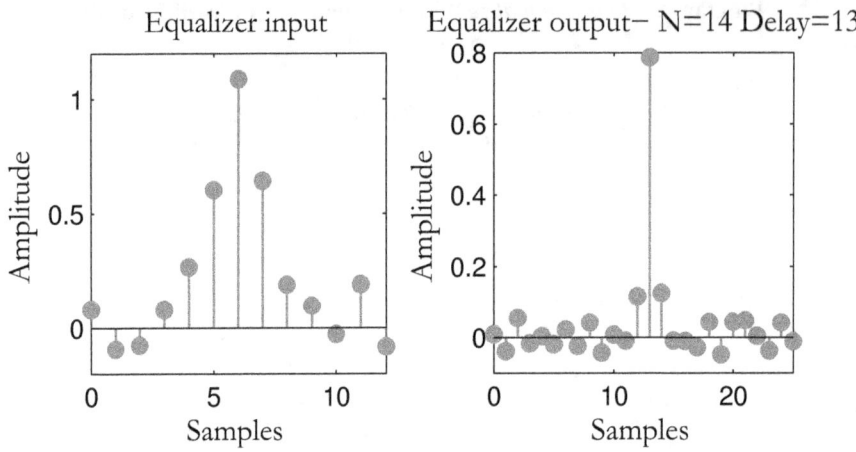

Fig. 5.11: MMSE equalizer of length $L = 14$ and optimized delay $n_0 = 13$

5.6 Equalizer delay optimization

For *preset* equalizers, the tap length N and the decision delay k_0 are fixed. The chosen tap length and equalizer delay, significantly affect the performance of the entire communication link. This gives rise to the need for optimizing the tap length and the decision delay. In this section, only the optimization of equalizer delay for both zero-forcing equalizer and MMSE equalizers are simulated. Readers may refer [8] for more details on this topic.

Equalizer delay is very crucial to the performance of the communication link. The equalizer delay determines the value of the symbol that is being detected at the current instant. Therefore, it directly affects the detector block that comes after the equalizer.

The Matlab functions, (shown in the previous sections), for designing zero forcing equalizer (*zf_equalizer.m*) and the MMSE equalizer (*mmse_equalizer.m*), already include the code that calculates the optimum decision delay. The optimization algorithm is based on equations 5.20 and 5.41.

The following code snippet demonstrates the need for delay optimization of an MMSE equalizer for each chosen equalizer tap lengths $N = [5, 10, 15, 20, 25, 30]$. For this simulation, the channel impulse response is chosen as $h[k] = [-0.1, -0.3, 0.4, 1, 0.4, 0.3, -0.1]$. The equalizer delay is swept for each case of tap length and the mean squared error is plotted for visual comparison (Figure 5.12). For a given tap length N, the equalizer with optimum delay is the one that gives the minimum MSE. The same methodology can be applied to test the delay-optimization for the zero forcing equalizer.

Program 5.10: *mmse_equalizer_delay_opti.m*: Delay optimization of MMSE equalizer for fixed N

```
1   %The test methodology takes the example given in
2   %Yu Gong et al., Adaptive MMSE equalizer with optimum tap-length and
3   %decision delay, sspd 2010 and try to get a similar plot as given in
4   %Figure 3 of the journal paper
5   h=[-0.1 -0.3 0.4 1 0.4 0.3 -0.1]; %test channel
6   SNR=10; %Signal-to-noise ratio at the equalizer input in dB
7   Ns=5:5:30; %sweep number of equalizer taps from 5 to 30
8   maxDelay=Ns(end)+length(h)-2; %max delay cannot exceed this value
9   mse=zeros(length(Ns),maxDelay); optimalDelay=zeros(length(Ns),1);
10
```

```
i=1;j=1;
for N=Ns, %sweep number of equalizer taps
  for delays=0:1:N+length(h)-2; %sweep delays
    %compute MSE and optimal delay for each combination
    [~,mse(i,j),optimalDelay(i,1)]=mmse_equalizer(h,SNR,N,delays);
    j=j+1;
  end
  i=i+1;j=1;
end
plot(0:1:N+length(h)-2,log10(mse.')) %plot mse in log scale
title('MSE Vs eq. delay for given channel and equalizer lengths');
xlabel('Equalizer delay');ylabel('log_{10}[mse]');
%display optimal delays for each selected filter length N. this will
%correspond with the bottom of the buckets displayed in the plot
disp('Optimal Delays for each N value ->');disp(optimalDelay);
```

Fig. 5.12: MSE Vs equalizer delay for different values of equalizer tap lengths N

5.7 BPSK Modulation with zero forcing and MMSE equalizers

Having designed and tested the zero forcing and MMSE FIR linear equalizers, the next obvious task is to test their performance over a communication link. The performance of the linear equalizers completely depend on the characteristics of the channel over which the communication happens. As an example, three distinct channels [4] with the channel characteristics shown in Figure 5.13, are put to test.

In the simulation model, given in Figure 5.14, a string of random data, modulated by BPSK modulation, is individually passed through the discrete channels given in 5.13. In addition to filtering the BPSK modulated

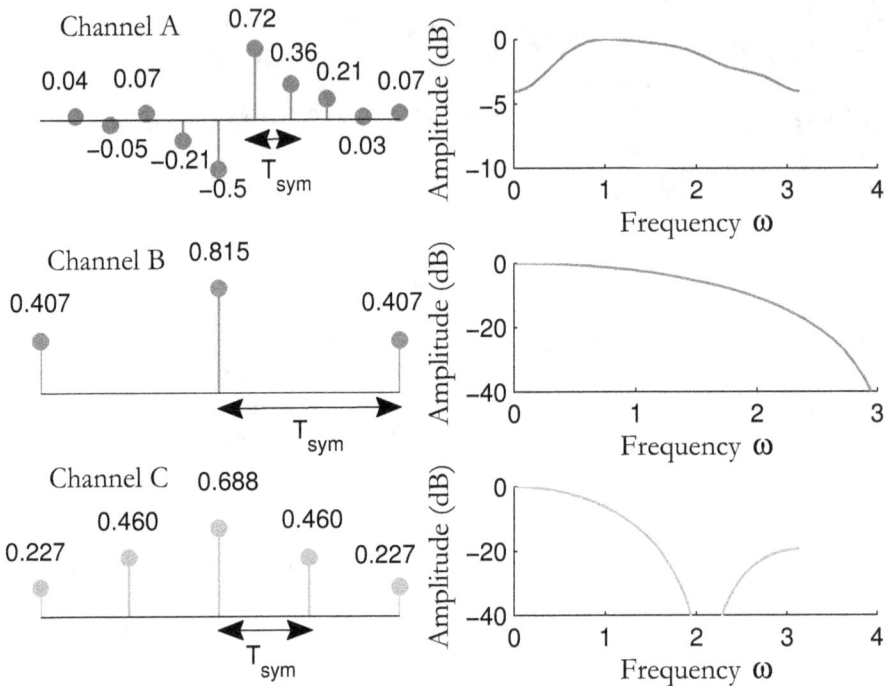

Fig. 5.13: Discrete time characteristics of three channels

data through the given channels, the transmitted data is also mixed with additive white Gaussian noise (as dictated by the given E_b/N_0 value). The received data is then passed through the *delay-optimized* versions of the zero-forcing equalizer and the MMSE equalizer, independently over two different paths. The outputs from the equalizers are passed through the threshold detector that gives an estimate of the transmitted information symbols. The detected symbols are then compared with the original data sequence and the bit-error-rate is calculated. The simulation is performed for different E_b/N_0 values and the resulting performance curves are plotted in Figure 5.15.

The time domain and frequency domain characteristics of the three channels A,B and C are plotted in Figure 5.13. The simulated performance results in Figure 5.15, also include the performance of the communication link over an ISI-free channel (no inter-symbol interference).

The performance of both the equalizers are the best in channel A, which has a typical response of a good telephone channel (channel response is gradual throughout the band with no spectral nulls). The MMSE equalizer performs poorly for channel C, which has the worst spectral characteristic. The next best performance of MMSE equalizer is achieved for channel B. Channel B still has spectral null but it is not as severe as that of channel C. The zero forcing filters perform the worst over both the channels B and C. This is because the design of zero forcing filter inherently omits the channel noise in the design equation for the computation of tap weights.

It can be concluded that the linear equalizers yield good performance over well-behaved channels that do not exhibit spectral nulls. On the other hand, they are inadequate for fully compensating inter-symbol interference over the channels that exhibit spectral nulls, which is often the reality. Decision feedback equalizers offer an effective solution for this problem. Non-linear equalizers with reduced complexity are also the focus of researchers to overcome the aforementioned limitations of linear equalizers.

Fig. 5.14: Simulation model for BPSK modulation with zero forcing and MMSE equalizers

Fig. 5.15: Error rate performance of MMSE and zero forcing FIR equalizers of length $N = 31$

The complete simulation code is given next. The simulation code reuses the add_awgn_noise.m function defined in section 4.1 of chapter 4 and the iqOptDetector.m function given in section 3.5.4 of chapter 3.

Program 5.11: *bpsk_isi_channels_equalizers.m*: Performance of FIR linear equalizers over ISI channels

```matlab
% Demonstration of Eb/N0 Vs SER for baseband BPSK modulation scheme
% over different ISI channels with MMSE and ZF equalizers
clear all; clc;
%---------Input Fields-----------------------
N=1e5;%Number of bits to transmit
EbN0dB = 0:2:30; % Eb/N0 range in dB for simulation
M=2; %2-PSK
h_c=[0.04 -0.05 0.07 -0.21 -0.5 0.72 0.36 0.21 0.03 0.07];%Channel A
h_c=[0.407 0.815 0.407]; %uncomment this for Channel B
h_c=[0.227 0.460 0.688 0.460 0.227]; %uncomment this for Channel C
nTaps = 31; %Desired number of taps for equalizer filter

SER_zf = zeros(length(EbN0dB),1); SER_mmse = zeros(length(EbN0dB),1);

%-----------------Transmitter--------------------
d= randi([0,1],1,N); %Uniformly distributed random source symbols
ref=cos(((M:-1:1)-1)/M*2*pi); %BPSK ideal constellation
s = ref(d+1); %BPSK Mapping
x = conv(s,h_c); %apply channel effect on transmitted symbols

for i=1:length(EbN0dB),
  %---------------Channel--------------------
  r=add_awgn_noise(x,EbN0dB(i));%add AWGN noise r = x+n

  %--------------Receiver--------------------
  %DELAY OPTIMIZED MMSE equalizer
  [h_mmse,MSE,optDelay]=mmse_equalizer(h_c,EbN0dB(i),nTaps);%MMSE eq.
  y_mmse=conv(h_mmse,r);%filter the received signal through MMSE eq.
  y_mmse=y_mmse(optDelay+1:optDelay+N);%samples from optDelay position

  %DELAY OPTIMIZED ZF equalizer
  [h_zf,error,optDelay]=zf_equalizer(h_c,nTaps);%design ZF equalizer
  y_zf=conv(h_zf,r);%filter the received signal through ZF equalizer
  y_zf = y_zf(optDelay+1:optDelay+N);%samples from optDelay position

  %Optimum Detection in the receiver - Euclidean distance Method
  %[estimatedTxSymbols,dcap]= iqOptDetector(y_zf,ref);%See chapter 3
  %Or a Simple threshold at zero will serve the purpose
  dcap_mmse=(y_mmse>=0);%1 is added since message symbols are m=1,2
  dcap_zf=(y_zf>=0);%1 is added since message symbols are m=1,2

  SER_mmse(i)=sum((d~=dcap_mmse))/N;%SER when filtered thro MMSE eq.
  SER_zf(i)=sum((d~=dcap_zf))/N;%SER when filtered thro ZF equalizer
end
theoreticalSER = 0.5*erfc(sqrt(10.^(EbN0dB/10)));%BPSK theory SER

figure;
semilogy(EbN0dB,SER_zf,'g'); hold on;
semilogy(EbN0dB,SER_mmse,'r','LineWidth',1.5);
```

```
50  semilogy(EbN0dB,theoreticalSER,'k'); ylim([1e-4,1]);
51  title('Probability of Symbol Error for BPSK signals');
52  xlabel('E_b/N_0 (dB)');ylabel('Probability of Symbol Error - P_s');
53  legend('ZF Equalizer','MMSE Equalizer','No interference');grid on;
54
55  [H_c,W]=freqz(h_c);%compute and plot channel characteristics
56  figure;subplot(1,2,1);stem(h_c);%time domain
57  subplot(1,2,2);plot(W,20*log10(abs(H_c)/max(abs(H_c))));%freq domain
```

5.8 Adaptive equalizer: Least mean square (LMS) algorithm

To solve for MMSE equalizer coefficients, the Weiner-Hopf solution (equation 5.36) requires the estimation of autocorrelation matrix $\mathbf{R}_{rr}$, correlation matrix $\mathbf{R}_{ra}$ and the matrix inversion $\mathbf{R}_{rr}^{-1}$. This method can be computationally intensive and may render potentially unstable filter.

Alternate solution is to use an iterative approach in which the filter weights $\mathbf{w}$ are updated iteratively in a direction towards the optimum Weiner solution ($\mathbf{R}_{rr}^{-1}\mathbf{R}_{ra}$). This is achieved by iteratively moving the filter weights $\mathbf{w}$ in the direction of the negative gradient

$$\mathbf{w}_{n+1} = \mathbf{w}_n - \mu \nabla \xi[n] \tag{5.42}$$

where, $\mathbf{w}_n$ is the vector of current filter tap weights, $\mathbf{w}_{n+1}$ is the vector of filter tap weights for the update, μ is the step size and $\nabla \xi[n]$ denotes the gradient of the mean square error as derived in equation 5.35.

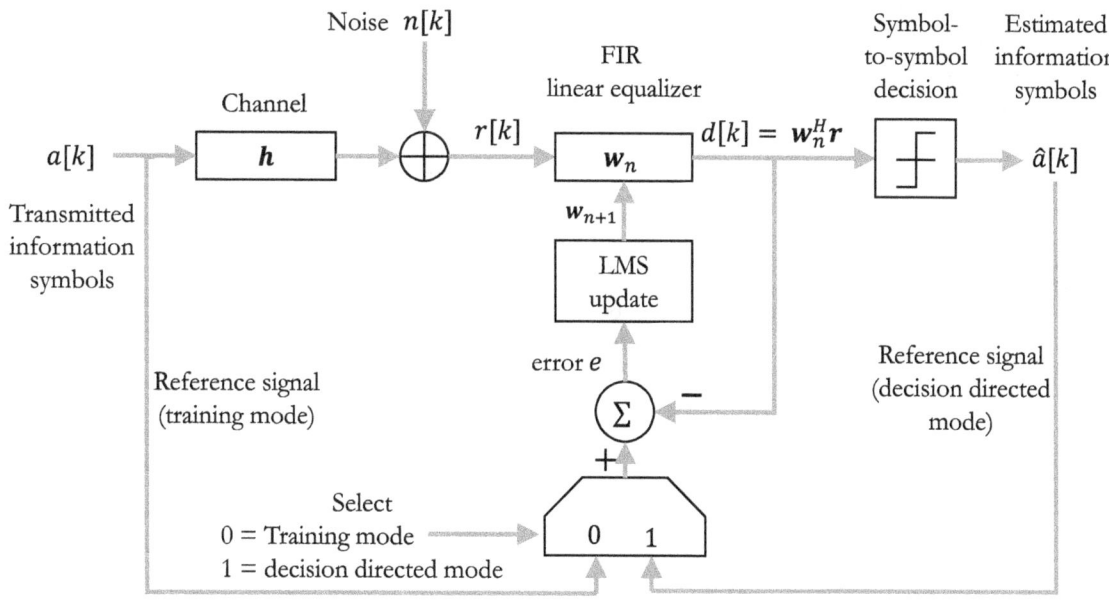

Fig. 5.16: Adaptive filter structure using LMS update

The *least mean square (LMS) algorithm* is derived from the gradient descent algorithm which is an iterative algorithm to search for the minimum error condition with the cost function equal to the mean square error at output of the equalizer filter. The algorithm is not realizable until we compute $\mathbf{R}_{rr} = E\left\{\mathbf{r}\mathbf{r}^H\right\}$ and $\mathbf{R}_{ra} = E\left\{\mathbf{r}a^*[k]\right\}$, involving an expectation operator. Therefore, for real-time applications, instantaneous squared error is preferred instead of mean squared error. This is achieved by replacing $\mathbf{R}_{rr} = E\left\{\mathbf{r}\mathbf{r}^H\right\}$ and $\mathbf{R}_{ra} = E\left\{\mathbf{r}a^*[k]\right\}$, with instantaneous estimates $\mathbf{R}_{rr} = \mathbf{r}\mathbf{r}^H$ and $\mathbf{R}_{ra} = \mathbf{r}a^*[k]$, respectively.

The channel model utilizing LMS algorithm for adapting the FIR linear equalizer is shown in Figure 5.16. The LMS algorithm computes the filter tap updates using the error between the reference sequence and the filter output. The channel model can be operated in two modes: training mode and decision directed mode. In training mode, the transmitted signal $a[k]$ acts as the reference signal for error generation. In decision directed mode (engaged when the actual data transmission happens), the output of the decision device $\hat{a}[k]$ acts as the reference signal for error generation.

With reference to the channel model in Figure 5.16, the recursive equations for computing the error term and the filter coefficient updates of the LMS algorithm in training mode are given by

$$e = a[k] - \mathbf{w}_n^H \mathbf{r} \qquad (5.43)$$

$$\mathbf{w}_{n+1} = \mathbf{w}_n + \mu e \mathbf{r} \qquad (5.44)$$

where, $\mathbf{w}_n = [w_0, w_1, \cdots, w_{N-1}]^H$ are the current filter coefficients at instant n with superscript H denoting Hermitian transpose and $\mathbf{r} = [r_k, r_{k-1}, \cdots, r_{k-N-1}]^T$ denotes the complex received sequence. For decision directed mode replace $a[k]$ with $\hat{a}[k]$ in the LMS equations.

The summary of the LMS algorithm for designing a filter of length N and its Matlab implementation are given.

Algorithm 1: Least mean square (LMS) algorithm

Input: N : filter length, μ: step size, r : received signal sequence, a: reference sequence

Output: w: final filter coefficients, e: estimation error

1 Initialization: $\mathbf{w}_n = zeros(N,1)$

2 **for** $k = N{:}length(r)$ **do**

3   Compute $\mathbf{r} = [r_k, r_{k-1}, \cdots, r_{k-N-1}]^T$

4   Compute error term $e = a[k] - \mathbf{w}_n^H \mathbf{r}$

5   Compute filter tap update $\mathbf{w}_{n+1} = \mathbf{w}_n + \mu e \mathbf{r}$

6 **return** w

Program 5.12: *lms.m*: Implementing LMS algorithm

```matlab
function w=lms(N,mu,r,a)
%Function implementing LMS update equations
% N = desired length of the filter
% mu = step size for the LMS update
% r = received/input signal
% a = reference signal
%
%Returns: % w = optimized filter coefficients

w=zeros(1,N);%initialize filter taps to zeros
for k=N:length(r)
    r_vector = r(k:-1:k-N+1);
    e = a(k)-w *r_vector';
    w = w + mu*e*r_vector;
end
end
```

To verify the implemented function, a simple test can be conducted. The adaptive filter should be able to identify the response of a short FIR filter whose impulse response is known prior to the test.

Program 5.13: *lms_test.m*: Verifying the LMS algorithm

```
1   %System identification using LMS algorithm
2   clearvars; clc;
3   N = 5; %length of the desired filter
4   mu=0.1; %step size for LMS algorithm
5
6   r=randn(1,10000);%random input signal
7   h=randn(1,N)+1i*randn(1,N); %random complex system
8   a=conv(h,r);%reference signal
9   w=lms(N,mu,r,a);%designed filter using input signal and reference
10
11  disp('System impulse response (h):'); disp(h)
12  disp('LMS adapted filter (w): '); disp(w);
```

A random input signal $r[k]$ and a random complex system impulse response are generated $h[n]$. They are convolved to produce the noiseless reference signal $a[k]$. Pretending that $h[n]$ is unknown, using only the input signal $r[k]$ and the reference signal $a[k]$, the LMS algorithm is invoked to design a FIR filter of length $N = 5$. The resulting optimized filter coefficients $w[n]$ should match the impulse response $h[n]$. Results from a sample run are given next.

```
System impulse response (h):
-0.5726 + 0.0607i 0.0361 - 0.5948i -1.1136 - 0.121i 0.5275 - 0.4212i 1.7004 + 1.8307i
LMS adapted filter (w):
-0.5726 + 0.0607i 0.0361 - 0.5948i -1.1136 - 0.121i 0.5275 - 0.4212i 1.7004 + 1.8307i
```

References

1. J. R. Treichler, I. Fijalkow, and C. R. Johnson, Jr., *Fractionally spaced equalizers: How long should they really be?*, IEEE Signal Processing Mag., vol. 13, pp. 65–81, May 1996
2. Edward A. Lee, David G. Messerschmitt John R. Barry, *Digital Communications*, 3rd ed.: Springer, 2004
3. R.W. Lucky, *Automatic equalization for digital communication*, Bell System Technical Journal 44, 1965
4. J. G. Proakis, *Digital Communications*, 3rd ed.: McGraw-Hill, 1995.
5. Monson H. Hayes , *Statistical Digital Signal Processing and Modeling*, chapter 4.4.5, Application FIR least squares inverse filter, Wiley, 1 edition, April 11, 1996.
6. C.R Johnson, Hr. et al., *On Fractionally-Spaced Equalizer Design for Digital Microwave Radio Channels*, Signals, Systems and Computers, 1995. 1995 Conference Record of the Twenty-Ninth Asilomar Conference on, vol. 1, pp. 290-294, November 1995.
7. Phil Schniter, *MMSE Equalizer Design*,March 6, 2008, http://www2.ece.ohio-state.edu/~schniter/ee501/handouts/mmse_eq.pdf
8. Yu Gong et al., *Adaptive MMSE Equalizer with Optimum Tap-length and Decision Delay*, Sensor Signal Processing for Defence (SSPD 2010), 29-30 Sept. 2010.

Chapter 6
Receiver Impairments and Compensation

Abstract IQ signal processing is widely used in today's communication receivers. All the IQ processing receiver structures suffer from problems due to amplitude and phase mismatches in their I and Q branches. IQ imbalances are commonplace in analog front-end circuitry and it results in interference of image signal on top of the desired signal. The focus of this chapter is to introduce the reader to some of the basic models of receiver impairments like phase imbalances, gain imbalances and DC offsets. Well known compensation techniques for DC offsets and IQ imbalances are also discussed with code implementation and results.

6.1 Introduction

Direct conversion image rejection receivers [1] for IQ processing - are quite popular in modern RF receiver front-end designs. Direct conversion receivers are preferred over conventional superheterodyne receivers because they do not require separate image filtering [2] [3]. In a direct conversion receiver, the received signal is amplified and filtered at baseband rather than at some intermediate frequency [2]. A direct conversion receiver shown in Figure 6.1 will completely reject the image bands, only if the following two conditions are met : 1) The local oscillator tuned to the desired RF frequency must produce the cosine and sine signals with a phase difference of exactly $90°$, 2) the gain and phase responses of the I and Q branches must match perfectly. In reality, analog components in the RF front-end are not perfect and hence it is impossible to satisfy these requirements. Therefore, complete rejection of the image bands, during RF-IQ conversion, is unavoidable.

The IQ imbalance results from non-ideal RF front-end components due to power imbalance and/or non-orthogonality of the I,Q branches caused by imperfect local oscillator outputs. In Figure 6.1, the power imbalance on the IQ branches is captured by the gain parameter g on the quadrature branch. The phase error between the local oscillator outputs is captured by the parameter ϕ.

The effect of IQ imbalance is quite disastrous for higher order modulations that form the basis for many modern day communication systems like IEEE 802.11 WLAN, UMTS, LTE, etc. Furthermore, the RF front-end may also introduce DC offsets in the IQ branches, leading to more performance degradation. In order to avoid expensive RF front-end components, signal processing algorithms for compensating the IQ imbalance and DC offsets are a necessity.

In this section, we begin by constructing a model, shown in Figure 6.2 , to represent the effect of following RF receiver impairments

- Phase imbalance and cross-talk on I,Q branches caused by local oscillator phase mismatch - ϕ.
- Gain imbalance on the I,Q branches - g.
- DC offsets in the I and Q branches - dc_i, dc_q.

In this model, the complex signal $r = r_i + jr_q$ denotes the perfect unimpaired signal to the receiver and $z = z_i + jz_q$ represents the signal after the introduction of IQ imbalance and DC offset effects in the receiver

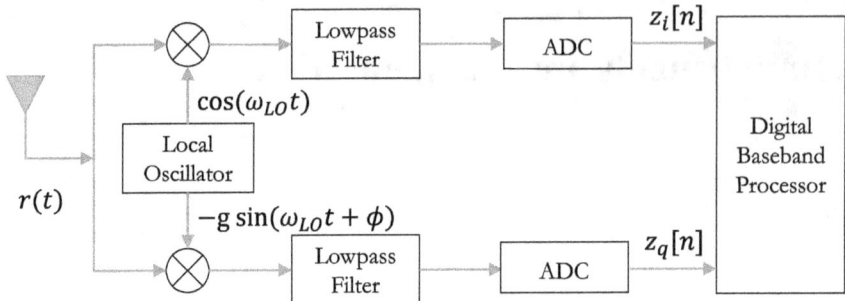

Fig. 6.1: Direct conversion image rejection receiver

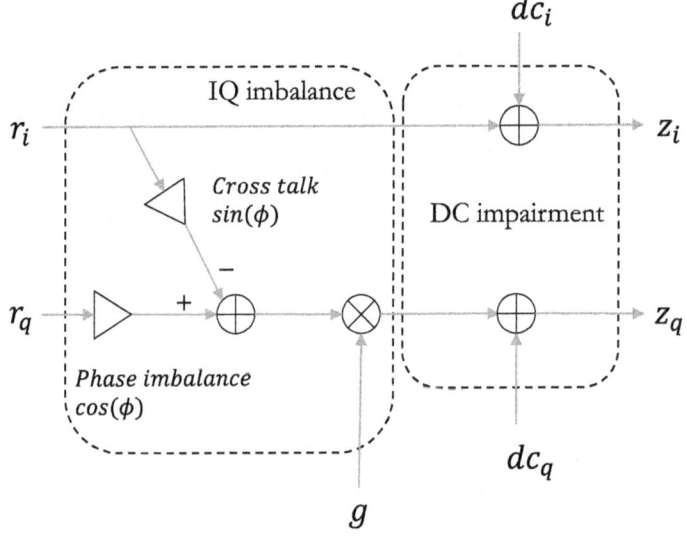

Fig. 6.2: RF receiver impairments model

front-end. The following RF receiver impairment model implemented in Matlab, consists of two steps. First it introduces IQ imbalance in the incoming complex signal and then adds DC offsets to the in-phase and quadrature branches. The details of the individual sub-functions for the IQ imbalance model and DC offset model are described in the subsequent sections. The DC offset impairment model and its compensation is discussed first, followed by the IQ impairment modeling and compensation.

Program 6.1: *receiver_impairments.m*: Function to add RF receiver impairments

```
function z=receiver_impairments(r,g,phi,dc_i,dc_q)
%Function to add receiver impairments to the IQ branches
%[z]=iq_imbalance(r,g,phi) introduces DC and IQ imbalances
% between the inphase and quadrature components of the complex
% baseband signal r. The model parameter g represent the gain
% mismatch between the IQ branches of the receiver and the
% parameter 'phi' represents the phase error of the local
% oscillator (in degrees). The DC biases associated with each
% I,Q path are represented by the paramters dc_i and dc_q.
```

```
10
11  k = iq_imbalance(r,g,phi); %Add IQ imbalance
12  z = dc_impairment(k,dc_i,dc_q); %Add DC impairment
```

6.2 DC offsets and compensation

The RF impairment model in the Figure 6.2 contains two types of impairments namely IQ imbalance and DC offset impairment. DC offsets dc_i and dc_q on the I and Q branches are simply modelled as additive factors on the incoming signal.

Program 6.2: *dc_impairment.m*: Model for introducing DC offsets in the IQ branches

```
1  function [y]=dc_impairment(x,dc_i,dc_q)
2  %Function to create DC impairments in a complex baseband model
3  %    [y]=iq_imbalance(x,dc_i,dc_q) introduces DC imbalance
4  %    between the inphase and quadrature components of the complex
5  %    baseband signal x. The DC biases associated with each I,Q path
6  %    are represented by the paramters dc_i and dc_q
7  y = x + (dc_i+1i*dc_q);
```

Correspondingly, the DC offsets on the branches are simply removed by subtracting the mean of the signal on the I,Q branches from the incoming signal.

Program 6.3: *dc_compensation.m*: Model for introducing DC offsets in the IQ branches

```
1  function [v]=dc_compensation(z)
2  %Function to estimate and remove DC impairments in the IQ branch
3  %    v=dc_compensation(z) removes the estimated DC impairment
4  iDCest=mean(real(z));%estimated DC on I branch
5  qDCest=mean(imag(z));%estimated DC on I branch
6  v=z-(iDCest+1i*qDCest);%remove estimated DCs
```

6.3 IQ imbalance model

Two IQ imbalance models, namely, *single-branch IQ imbalance model* and *double-branch IQ imbalance model*, are used in practice. These two models differ in the way the IQ mismatch in the inphase and quadrature arms are envisaged. In the single-branch model, the IQ mismatch is modeled as amplitude and phase errors in only one of the branches (say Q branch). In contrast, in the double-branch model, the IQ mismatch is modeled as amplitude and phase errors in both the I and Q branches. The estimation and compensation algorithms for both these models have to be adjusted accordingly. In this text, only the single-branch IQ imbalance model and its corresponding compensation algorithms are described.

With reference to Figure 6.2, let $r = r_i + jr_q$ denote the perfect unimpaired signal and $z = z_i + jz_q$ represent the signal with IQ imbalance. In this single-branch IQ imbalance model, the gain mismatch is represented by a gain term g in the Q branch only. The difference $1 - g$ is a measure of amplitude deviation of Q branch from the perfectly balanced condition. The presence of phase error π in the local oscillator outputs, manifests as cross-talk between the I and Q branches. In essence, together with the gain imbalance g and the phase error ϕ, the impaired signals on the I,Q branches are represented as

$$\begin{bmatrix} z_i[k] \\ z_q[k] \end{bmatrix} = \begin{bmatrix} 1 & 0 \\ -g.sin(\phi_{rad}) & g.cos(\phi_{rad}) \end{bmatrix} \begin{bmatrix} r_i[k] \\ r_q[k] \end{bmatrix} \tag{6.1}$$

For the given gain mismatch g and phase error ϕ, the following Matlab function introduces IQ imbalance in a complex baseband signal.

Program 6.4: *iq_imbalance.m*: IQ imbalance model

```
1  function [z]= iq_imbalance(r,g,phi)
2  %Function to create IQ imbalance impairment in a complex baseband
3  % [z]=iq_imbalance(r,g,phi) introduces IQ imbalance and phase error
4  % signal between the inphase and quadrature components of the
5  % complex baseband signal r. The model parameter g represents the
6  % gain mismatch between the IQ branches of the receiver and 'phi'
7  % represents the phase error of the local oscillator (in degrees).
8  Ri=real(r); Rq=imag(r);
9  Zi= Ri; %I branch
10 Zq= g*(-sin(phi/180*pi)*Ri + cos(phi/180*pi)*Rq);%Q branch crosstalk
11 z=Zi+1i*Zq;
12 end
```

6.4 IQ imbalance estimation and compensation

Several IQ imbalance compensation schemes are available. This section uses two of them which are based on the works of [4] and [5]. The first algorithm [4] is a blind estimation and compensation algorithm and the one described in [5] is a pilot based estimation and compensation algorithm. Both these techniques are derived for the single-branch IQ imbalance model.

6.4.1 Blind estimation and compensation

The blind technique is a simple, low complexity algorithm, that is solely based on the statistical properties of the incoming complex signal. It does not require additional processing overheads like preamble or training symbols for the IQ imbalance estimation.

The blind compensation scheme is shown in Figure 6.3. Let $z = z_i + jz_q$, represent the IQ imbalance-impaired complex baseband signal, that needs to be compensated. First, the complex baseband signal z is used to estimate three imbalance parameters θ_1, θ_2 and θ_3.

$$\theta_1 = -E\left\{sgn(z_i)z_q\right\}$$
$$\theta_2 = E\left\{|z_i|\right\}$$
$$\theta_3 = E\left\{|z_q|\right\} \tag{6.2}$$

where, $sgn(x)$ is the signum function

$$sgn(x) = \begin{cases} -1 & x < 0 \\ 0 & x = 0 \\ 1 & x > 0 \end{cases} \tag{6.3}$$

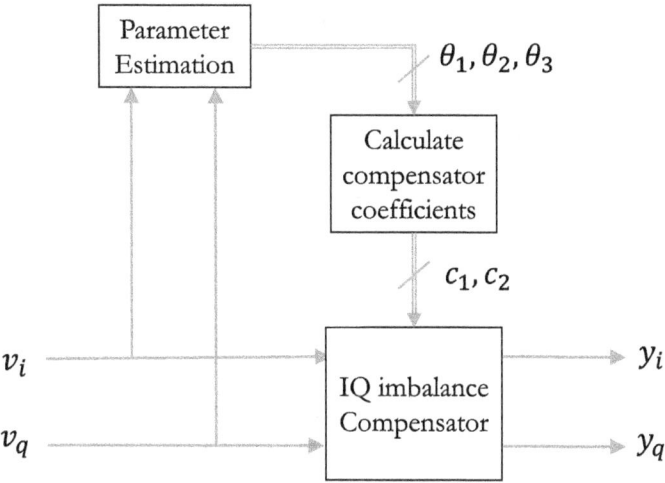

Fig. 6.3: Blind estimation and compensation for IQ imbalance

The compensator co-efficients c_1 and c_2 are then calculated from the estimates of θ_1, θ_2 and θ_3.

$$c_1 = \frac{\theta_1}{\theta_2} \qquad c_2 = \sqrt{\frac{\theta_3^2 - \theta_1^2}{\theta_2^2}} \qquad (6.4)$$

Finally, the inphase and quadrature components of the compensated signal $y = y_i + jy_q$, are calculated as

$$y_i = z_i \qquad y_q = \frac{c_1 z_i + z_q}{c_2}$$

Program 6.5: *blind_iq_compensation.m*: Blind IQ imbalance estimation and compensation

```
1  function y=blind_iq_compensation(z)
2  %Function to estimate and compensate IQ impairments for the single-
3  %branch IQ impairment model
4  % y=blind_iq_compensation(z) estimates and compensates IQ imbalance
5  I=real(z);Q=imag(z);
6  theta1=(-1)*mean(sign(I).*Q);
7  theta2=mean(abs(I));
8  theta3=mean(abs(Q));
9  c1=theta1/theta2; c2=sqrt((theta3^2-theta1^2)/theta2^2);
10 yI = I; yQ = (c1*I+Q)/c2; y= (yI +1i*yQ);
11 end
```

6.4.2 Pilot based estimation and compensation

Pilot based estimation algorithms are widely used to estimate various channel properties. The transmitter transmits a pilot sequence and the IQ imbalance is estimated based on the received sequence at the baseband signal processor. The pilot estimation technique is well suited for wireless applications like WLAN, UMTS and LTE, where a known pilot sequence is transmitted as part of the data communication session.

The low complexity algorithm, proposed in [5], uses a preamble of length $L = 64$ in order to estimate the gain imbalance (K_{est}) and phase error P_{est}.

$$K_{est} = \sqrt{\frac{\sum_{k=1}^{L} z_q^2[k]}{\sum_{k=1}^{L} z_i^2[k]}} \qquad (6.5)$$

$$P_{est} = \frac{\sum_{k=1}^{L} (z_i[k].z_q[k])}{\sum_{k=1}^{L} z_i^2[k]} \qquad (6.6)$$

where, the complex signal $z = z_i + jz_q$, represents the impaired version of the long preamble as given in the IEEE 802.11a specification [6].

Program 6.6: *pilot_iq_imb_est.m*: IQ imbalance estimation using Pilot transmission

```
1   function [Kest,Pest]=pilot_iq_imb_est(g,phi,dc_i,dc_q)
2   %Length 64 - Long Preamble as defined in the IEEE 802.11a
3   preamble_freqDomain = [0,0,0,0,0,0,1,1,-1,-1,1,1,-1,1,...
4       -1,1,1,1,1,1,1,-1,-1,1,1,-1,1,-1,1,1,1,1,1,...
5       0,1,-1,-1,1,1,-1,1,-1,1,-1,-1,-1,-1,-1,1,1,...
6       -1,-1,1,-1,1,-1,1,1,1,1,0,0,0,0,0];%freq. domain representation
7   preamble=ifft(preamble_freqDomain,64);%time domain representation
8
9   %send known preamble through DC & IQ imbalance model and estimate it
10  r=receiver_impairments(preamble,g,phi,dc_i,dc_q);
11  z=dc_compensation(r); %remove DC imb. before IQ imbalance estimation
12  %IQ imbalance estimation
13  I=real(z); Q=imag(z);
14  Kest = sqrt(sum((Q.*Q))./sum(I.*I)); %estimate gain imbalance
15  Pest = sum(I.*Q)./sum(I.*I); %estimate phase mismatch
16  end
```

Let $d = d_i + jd_q$ be the impaired version of complex signal received during the normal data transmission interval. Once the parameters given in equation 6.5 and 6.6 are estimated during the preamble transmission, the IQ compensation during the normal data transmission is as follows

$$w_i = d_i$$
$$w_q = \frac{d_q - P_{est}.d_i}{k_{est}\sqrt{1 - P_{est}^2}} \qquad (6.7)$$

Program 6.7: *iqImb_compensation.m*: Compensation of IQ imbalance during data transmission interval

```
1   function y=iqImb_compensation(d,Kest,Pest)
2   %Function to compensate IQ imbalance during the data transmission
3   % y=iqImb_compensation(d,Kest,Pest) compensates the IQ imbalance
4   % present at the received complex signal d at the baseband
5   % processor. The IQ compensation is performed using the gain
6   % imbalance (Kest) and phase error (Pest) parameters that are
7   % estimated during the preamble transmission.
8   I=real(d); Q=imag(d);
9   wi= I;
10  wq = (Q - Pest*I)/sqrt(1-Pest^2)/Kest;
11  y = wi + 1i*wq;
12  end
```

6.5 Visualizing the effect of receiver impairments

With the receiver impairments model and compensation techniques in place, let us visualize their effects in a complex plane. For instance, consider a higher order modulation like 64-QAM. The receiver impairments like IQ imbalance and DC offsets are added to the QAM modulated signal. The unimpaired sequence and the sequence affected by receiver impairments are plotted on a complex plane as shown in Figure 6.4. The following code re-uses the mqam_modulator function that was already defined in section 3.4.3 of chapter 3.

Program 6.8: *test_rf_impairments.m*: Visualize receiver impairments in a complex plane

```
1   clearvars; clc;
2   M=64;%M-QAM modulation order
3   N=1000;%To generate random symbols
4   d=ceil(M.*rand(N,1));%random data symbol generation
5   s = mqam_modulator(M,d); %M-QAM modulated symbols (s)
6
7   g_1=0.8; phi_1=0; dc_i_1=0; dc_q_1=0; %gain mismatch only
8   g_2=1;   phi_2=12; dc_i_2=0; dc_q_2=0; %phase mismatch only
9   g_3=1;   phi_3=0; dc_i_3=0.5; dc_q_3=0.5; %DC offsets only
10  g_4=0.8; phi_4=12; dc_i_4=0.5; dc_q_4=0.5; %All impairments
11
12  r1=receiver_impairments(s,g_1,phi_1,dc_i_1,dc_q_1);
13  r2=receiver_impairments(s,g_2,phi_2,dc_i_2,dc_q_2);
14  r3=receiver_impairments(s,g_3,phi_3,dc_i_3,dc_q_3);
15  r4=receiver_impairments(s,g_4,phi_4,dc_i_4,dc_q_4);
16
17  subplot(2,2,1);plot(real(s),imag(s),'b.');hold on;
18  plot(real(r1),imag(r1),'r.'); title('IQ Gain mismatch only')
19  subplot(2,2,2);plot(real(s),imag(s),'b.');hold on;
20  plot(real(r2),imag(r2),'r.'); title('IQ Phase mismatch only')
21  subplot(2,2,3);plot(real(s),imag(s),'b.');hold on;
22  plot(real(r3),imag(r3),'r.'); title('DC offsets only')
23  subplot(2,2,4);plot(real(s),imag(s),'b.');hold on;
24  plot(real(r4),imag(r4),'r.');title('IQ impairments & DC offsets');
```

Fig. 6.4: Constellation plots for unimpaired (blue) and impaired (red) versions of 64-QAM modulated symbols: (a) Gain imbalance $g = 0.8$ (b) Phase mismatch $\phi = 12°$ (c) DC offsets $dc_i = 0.5, dc_q = 0.5$ (d) With all impairments $g = 0.8, \phi = 12°, dc_i = 0.5, dc_q = 0.5$

6.6 Performance of M-QAM modulation with receiver impairments

In any physical layer of a wireless communication system that uses OFDM with higher order M-QAM modulation, the effect of receiver impairments is of great concern. The performance of a higher order M-QAM under different conditions of receiver impairments is simulated here. Figure 6.5 shows the functional blocks used in this performance simulation. A received M-QAM signal is impaired by IQ imbalance (both gain imbalance g and phase imbalance ϕ) and DC offsets (dc_i, dc_q) on the IQ arms. First, the DC offset is estimated and compensated; Next, the IQ compensation is separately performed using blind IQ compensation and pilot based IQ compensation. The symbol error rates are finally computed for following four cases: (a) received signal (with no compensation), (b) with DC compensation alone, (c) with DC compensation and blind IQ compensation and (d) with DC compensation and pilot based IQ compensation.

The complete simulation code for this performance simulation is given next. The code re-uses many functions defined in chapter 3 and chapter 4. The function to perform MQAM modulation and the coherent detection technique was already defined in sections 3.4.3 and 3.5.3. The AWGN noise model and the code to compute theoretical symbol error rates are given in sections 4.1.2 and 4.1.3 respectively.

Figure 6.6 illustrates the results obtained for the following values of receiver impairments $g = 0.9, \phi = 8°, dc_i = 1.9, dc_q = 1.7$. For this condition, both the blind IQ compensation and pilot based IQ compensation techniques are on par. Figure 6.7 illustrates the results obtained for the following values of receiver impairments $g = 0.9, \phi = 30°, dc_i = 1.9, dc_q = 1.7$, where the phase mismatch is drastically increased compared to the previous case. For this condition, the performance with the pilot based IQ compensation is better than that of the blind compensation technique. Additionally, the constellation plots of the impaired signal and the compensated signal, are also plotted for various combinations of DC offsets, gain imbalances and phase imbalances, which is shown in Figure 6.8.

Program 6.9: *mqam_iq_imbalance_awgn.m*: Performance of M-QAM with receiver impairments

```matlab
%Eb/N0 Vs SER for M-QAM modulation with receiver impairments
clear all;clc;
%---------Input Fields-----------------------
N=100000; %Number of input symbols
EbN0dB = -4:5:24; %Define EbN0dB range for simulation
M=64; %M-QAM modulation order
g=0.9; phi=8; dc_i=1.9; dc_q=1.7;%receiver impairments
%----------------------------------------------
k=log2(M); %Bits per symbol
EsN0dB = 10*log10(k)+EbN0dB; %Converting Eb/N0 to Es/N0
SER1 = zeros(length(EsN0dB),1);%Symbol Error rates (No compensation)
SER2 = SER1;%Symbol Error rates (DC compensation only)
SER3 = SER1;%Symbol Error rates (DC comp & Blind IQ compensation)
SER4 = SER1;%Symbol Error rates (DC comp & Pilot IQ compensation)

d=ceil(M.*rand(1,N));%random data symbols drawn from [1,2,..,M]
[s,ref]=mqam_modulator(M,d);%MQAM symbols & reference constellation
%See section 3.4 on chapter 'Digital Modulators and Demodulators
%- Complex Baseband Equivalent Models' for function definition

for i=1:length(EsN0dB),
 r = add_awgn_noise(s,EsN0dB(i)); %see section 4.1 on chapter 4

 z=receiver_impairments(r,g,phi,dc_i,dc_q);%add impairments

 v=dc_compensation(z); %DC compensation
 y3=blind_iq_compensation(v); %blind IQ compensation
 [Kest,Pest]=pilot_iq_imb_est(g,phi,dc_i,dc_q);%Pilot based estimation
 y4=iqImb_compensation(v,Kest,Pest);%IQ comp. using estimated values

 %Enable this section - if you want to plot constellation diagram -
 %figure(1);plot(real(z),imag(z),'rO'); hold on;
 %plot(real(y4),imag(y4),'b*'); hold off;
 %title(['Eb/N0=',num2str(EbN0dB(j)),' (dB)']);pause;

 %-------IQ Detectors - defined in section 3.5.4 chapter 3--------
 [estTxSymbols_1,dcap_1]= iqOptDetector(z,ref);%No compensation
 [estTxSymbols_2,dcap_2]= iqOptDetector(v,ref);%DC compensation only
 [estTxSymbols_3,dcap_3]= iqOptDetector(y3,ref);%DC & blind IQ comp.
 [estTxSymbols_4,dcap_4]= iqOptDetector(y4,ref);%DC & pilot IQ comp.
 %------ Symbol Error Rate Computation-------
 SER1(i)=sum((d~=dcap_1))/N; SER2(i)=sum((d~=dcap_2))/N;
 SER3(i)=sum((d~=dcap_3))/N; SER4(i)=sum((d~=dcap_4))/N;
end
theoreticalSER = ser_awgn(EbN0dB,'MQAM',M); %theoretical SER
figure(2);semilogy(EbN0dB,SER1,'r*-'); hold on; %Plot results
semilogy(EbN0dB,SER2,'bO-'); semilogy(EbN0dB,SER3,'g^-');
semilogy(EbN0dB,SER4,'m*-'); semilogy(EbN0dB,theoreticalSER,'k');
legend('No compensation','DC comp only',...
```

```
50   'Sim- DC & blind iq comp','Sim- DC & pilot iq comp','Theoretical');
51   xlabel('E_b/N_0 (dB)');ylabel('Symbol Error Rate (Ps)');
52   title('Probability of Symbol Error 64-QAM signals');
```

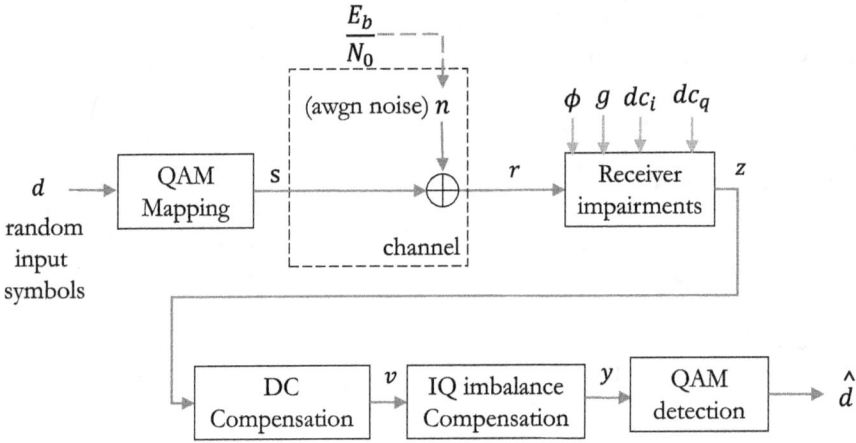

Fig. 6.5: Simulation model for M-QAM system with receiver impairments

Fig. 6.6: Performance of 64-QAM with receiver impairments: $g = 0.9, \phi = 8°, dc_i = 1.9, dc_q = 1.7$

Fig. 6.7: Performance of 64-QAM with receiver impairments: $g = 0.9, \phi = 30°, dc_i = 1.9, dc_q = 1.7$

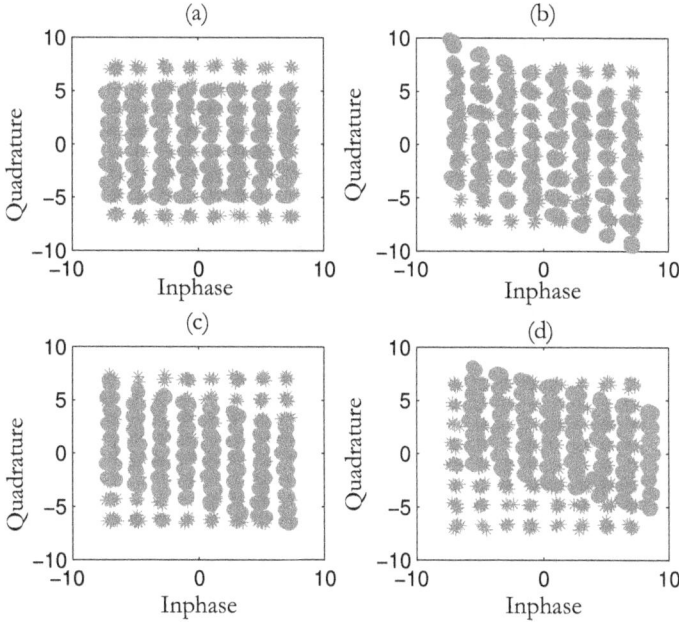

Fig. 6.8: Constellation plots at $E_b/N_0 = 21dB$ for various combinations of receiver impairments: (a) Gain imbalance only - $g = 0.7$ (b) Phase mismatch only - $\phi = 25°$ (c) Gain imbalance and phase mismatch $g = 0.7, \phi = 25°$ (d) All impairments - $g = 0.7, \phi = 25°, dc_i = 1.5, dc_q = 1.5$

References

1. T.H. Meng W. Namgoong, *Direct-conversion RF receiver design*, IEEE Transaction on Communications, 49(3):518–529, March 2001
2. A. Abidi, *Direct-conversion radio transceivers for digital communications*, IEEE Journal of Solid-State Circuits, 30:1399–1410, December 1995.

3. B. Razavi, RF microelectronics, ISBN 978-0137134731, Prentice Hall, 2 edition, October 2011
4. Moseley, Niels A., and Cornelis H. Slump. *A low-complexity feed-forward I/Q imbalance compensation algorithm*, 17th Annual Workshop on Circuits, 23-24 Nov 2006, Veldhoven, The Netherlands. pp. 158-164. Technology Foundation STW. ISBN 978-90-73461-44-4
5. K.H Lin et al., *Implementation of Digital IQ Imbalance Compensation in OFDM WLAN Receivers*, IEEE International Symposium on Circuits and Systems, 2006
6. *Wireless LAN Medium Access Control (MAC) and Physical Layer (PHY) Specifications: High-Speed Physical Layer in the 5 GHz Band*, IEEE Standard 802.11a-1999-Part II, Sep.1999.

Index

www.ingramcontent.com/pod-product-compliance
Lightning Source LLC
Chambersburg PA
CBHW081443170526
45166CB00008B/2296